Y0-BYB-295

HMO ENERGY
CHARACTERISTICS

HMO ENERGY CHARACTERISTICS

Rudolf Zahradník and Jiří Pancíř

Institute of Physical Chemistry
Czechoslovak Academy of Sciences
Prague, Czechoslovakia

QD461
Z19h
1970

IFI/PLENUM • NEW YORK-WASHINGTON-LONDON • 1970

258915

Library of Congress Catalog Card Number 75-130314

SBN 306-65152-1

© 1970   IFI/Plenum Data Corporation
A Subsidiary of Plenum Publishing Corporation
227 West 17th Street, New York, N. Y. 10011

United Kingdom edition published by Plenum Press, London
A Division of Plenum Publishing Company, Ltd.
Donington House, 30 Norfolk Street, London W.C.2, England

All rights reserved

No part of this publication may be reproduced in any form
without written permission from the publisher

Printed in the United States of America

CONTENTS

INTRODUCTION

The present collection of HMO data, in contrast to large collections published recently,[1,2] includes *only* HMO *energy characteristics* (see also Refs. 3–5). Clearly, the field of applicability of these data is narrower; but on the other hand, it makes it possible to collect data for about 750 systems including numerous large series of vinylogs and benzologs of various parent systems. The data concern hydrocarbons and hydrocarbon ions with delocalized charges; these systems were dealt with using the classical Hückel version of the MO-LCAO method.

Values of the following characteristics are tabulated: number of the system, its chemical name, number of π-electrons (n), number of σ-bonds between conjugated atoms (m), π-electronic energy (W), specific delocalization energy (DE_{sp}), coefficients in the expressions for the energy of frontier MO (k_2, k_1, k_{-1}, k_{-2}), and energies for the $N \to V_1$ transitions [$E(N \to V_1)$]. Well known expressions were used for calculation of these characteristics:

$$W = \sum_i n_i E_i,$$

where E_i stands for the energy of the ith MO and n_i is the number of electrons in this orbital (0, 1, 2); for the energies of the MO, the following expression holds:

$$E_i = a + k_i \beta,$$

where a and β stand for the Coulomb and resonance integrals of the HMO method, respectively. The subscripts 1 and 2 denote the highest occupied and the next nearest MO, the subscripts -1 and -2 denote similarly the lowest free MO's. Further, we have

$$E(N \to V_1) = |k_{-1} - k_1|,$$

and, finally,

$$DE_{sp} = \frac{W - W_L}{m},$$

where W_L means the HMO energy of the Kekulé (localized) structure. All energy characteristics are given in β-units. The contributions to the π-electron energy expressed in terms of a are neglected for the sake of simplicity so that in fact the binding energies are tabulated.

1

The characteristics included in this collection can be used to estimate[6-8] π-electronic energy, aromaticity, ionization potentials, electron affinities (oxidizability and reducibility), polarographic half-wave potentials, wave numbers of some bands in electronic spectra of individual compounds (this concerns even radicals[9]), and charge-transfer complexes.

The groups of compounds on the following pages are denoted by a two-digit code; three additional digits are used to identify the compounds belonging to these groups. Thus, each system is described by a set of five numbers. In order to make the orientation easier, these numbers appear at the top of the pages in both the formula and the numerical parts of these tables.

In general the names of systems included in *Revised Ring Index* are used. The names of other systems are formed along the lines suggested in the *Revised Ring Index*. Of course these names have to be treated as tentative names only.

The data included in this collection originate from papers of our group (Refs. 10–20) and from papers of co-workers of Prof. J. Koutecký (Refs. 21–25). All these data were checked by new calculations using an HMO program designed by J. Pancíř for the Elliott E 503 computer.

The accuracy of the energy characteristics [W, k_i, $E(N \rightarrow V_1)$] amounts to $\pm 10^{-5}$ β-units.

REFERENCES

1. Heilbronner, E., and Straub, P. A. *Hückel Molecular Orbitals,* Springer-Verlag, Berlin, 1966.
2. Streitwieser, A., Jr., and Brauman, J. I. *Supplemental Tables of MO Calculations,* Pergamon Press, Oxford, 1965.
3. Coulson, C. A., and Daudel, R. *Dictionary of Values of Molecular Constants (Wave mechanical methods),* The Mathematical Institute, Oxford and CNRS, Paris.
4. Pullman, B., and Pullman, A. *Quantum Biochemistry,* Interscience Publishers, New York, 1963.
5. Pullman, B., and Pullman, A. *Les théories électroniques de la chimie organique,* Masson, Paris, 1952.
6. Streitwieser, A., Jr. *Molecular Orbital Theory for Organic Chemists,* John Wiley & Sons, New York, 1961.
7. Heilbronner, E., and Bock, H. *Das HMO-Modell und seine Anwendung,* Verlag-Chemie, Weinheim, 1968.
8. Zahradník, R. *Fortschr. chem. Forsch.* **10,** 1 (1968).
9. Zahradník, R., and Čársky, P. *J. Phys. Chem.* **74,** 1240 (1970).
10. Zahradník, R., Michl, J., and Koutecký, J. *Collection Czechoslov. Chem. Commun.* **29,** 1932 (1964).
11. Zahradník, R., and Michl, J. *Collection Czechoslov. Chem. Commun.* **30,** 520 (1965).
12. Zahradník, R., and Michl, J. *Collection Czechoslov. Chem. Commun.* **30,** 1060 (1965).
13. Zahradník, R., Michl, J., and Pancíř, J. *Collection Czechoslov. Chem. Commun.* **30,** 2891 (1965).
14. Zahradník, R., and Párkányi, C. *Collection Czechoslov. Chem. Commun.* **30,** 3536 (1965).
15. Zahradník, R., Michl, J., and Jutz, C. *Collection Czechoslov. Chem. Commun.* **30,** 3227 (1965).
16. Michl, J. and Zahradník, R. *Collection Czechoslov. Chem. Commun.* **31,** 1475 (1966).
17. Zahradník, R., Michl, J., and Pancíř, J. *Tetrahedron* **22,** 1355 (1966).
18. Zahradník, R., and Michl, J. *Collection Czechoslov. Chem. Commun.* **31,** 3442 (1966).
19. Tichý, M., and Zahradník, R. *Collection Czechoslov. Chem. Commun.* **32,** 4485 (1967).
20. Kopecký, J. Unpublished results, 1966.
21. Hochmann, P., Dubský, J., Koutecký, J., and Párkányi, C. *Collection Czechoslov. Chem. Commun.* **30,** 3560 (1965).
22. Hochmann, P., Dubský, J., Kvasnička, V., and Titz, M. *Collection Czechoslov. Chem. Commun.* **31,** 4172 (1966).
23. Hochmann, P., Dubsky, J., and Titz, M. *Collection Czechoslov. Chem. Commun.* **32,** 1260 (1967).
24. Titz, M., and Hochmann, P. *Collection Czechoslov Chem. Commun.* **32,** 2343 (1967).
25. Titz, M., and Hochmann, P. *Collection Czechoslov. Chem. Commun.* **32,** 3028 (1967).

3

SYMBOLS

No.	number of a compound within the respective group
n	number of π-electrons
m	number of $C - C$ σ-bonds of the conjugated system
W	π-electronic energy of the ground state
DE_{sp}	specific delocalization energy
k_2, k_1, k_{-1} k_{-2}	coefficients in expressions for the energy of the highest occupied and lowest free MO's
$E(N \rightarrow V_1)$	energy of $N \rightarrow V_1$ transition
A	anion
diA	dianion
C	cation
diC	dication

KEY TO CODE DESIGNATIONS

11. Odd Polyenes   11001 − 11011

$$C \text{—(} C \text{—} C)_i$$
11

i = 1, 2, 3, 4, 5, 6, 7, 8, 9, 12, 17
No.1 2 3 4 5 6 7 8 9 10 11

12. Even Polyenes   12001 − 12014

$$C \text{—(} C \text{—} C \text{—)}_i C$$
12

i = 0, 1, 2, 3, 4, 5, 6, 7, 8, 9, 11, 12, 16, 17
No.1 2 3 4 5 6 7 8 9 10 11 12 13 14

13. Odd Cyclopolyenes   13001 − 13011

$$\begin{array}{c} C \\ / \ \backslash \\ (C \text{—} C)_i \end{array}$$
13

i = 1, 2, 3, 4, 5, 6, 7, 8, 9, 12, 17
No. 1 2 3 4 5 6 7 8 9 10 11

14. Even Cyclopolyenes   14001 − 14011

$$\begin{array}{c} C \text{—} C \\ | \quad | \\ (C \text{—} C)_i \end{array}$$
14

i = 1, 2, 3, 4, 5, 6, 7, 8, 9, 12, 17
No. 1 2 3 4 5 6 7 8 9 10 11

15. Fulvenes   15001 − 15005

$$\begin{array}{c} C \\ | \\ C \\ / \ \backslash \\ (C \text{—} C)_i \end{array}$$

i = 1, 2, 3, 4, 5
No. 1 2 3 4 5

16. Fulvalenes   16001 − 16015

i = 1, 1, 1, 1, 1, 2, 2, 2, 2
j = 1, 2, 3, 4, 5, 2, 3, 4, 5
No. 1 2 3 4 5 6 7 8 9

i = 3, 3, 3, 4, 4, 5
j = 3, 4, 5, 4, 5, 5
No. 10 11 12 13 14 15

17. Radialenes   17001 − 17009

i = 1, 2, 3, 4, 5, 6, 7, 8, 9
No. 1 2 3 4 5 6 7 8 9

21. Benzocyclobutadienes   21001-21008

1   2   3   4   5

  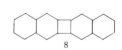

6   7   8

31. Benzenoid Hydrocarbons   31001-31115

(Even Systems)

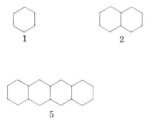

1   2   3   4

5   6   7

31. Benzenoid Hydrocarbons (Even Systems) (continued)

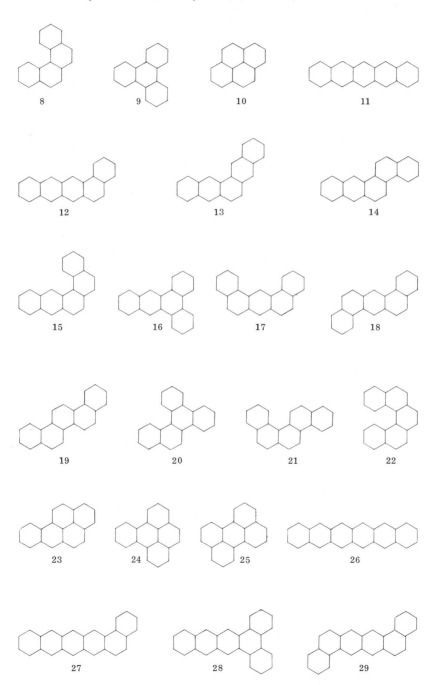

31. Benzenoid Hydrocarbons (Even Systems)   (continued)

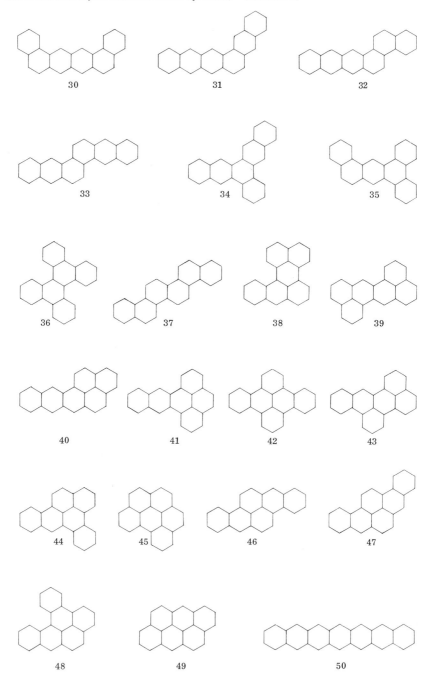

31. Benzenoid Hydrocarbons (Even Systems) (continued)

51

52

53

54

55

56

57

58

59

60

61

62

63

64

65

66

67

68

31. Benzenoid Hydrocarbons (Even Systems) (continued)

69

70

71

72

73

74

75

76

77

78

79

80

81

82

83

84

85

86

31. Benzenoid Hydrocarbons (Even Systems)   (continued)

87

88

89

90

91

92

93

94

95

96

97

98

99

100

101

102

103

31. Benzenoid Hydrocarbons (continued)

(Even Systems)

104

105

106

(Odd Systems)

107

108

109

110

111

112

113

114

115

32. Arylmethyl Ions

32. Arylmethyl Ions (continued)

Note: The integers in the formulas denote the positions of the exocyclic carbons and the numbers of the corresponding compounds in the tables.

33. Vinylpolyacenes   33001-33011

33. Vinylpolyacenes (continued)

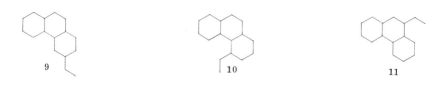

34. Quinodimethanes

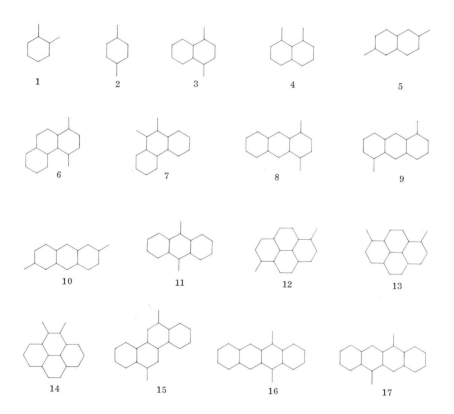

35. Phenylpolyenes

Even Systems

i = 1, 2, 3, 4, 5, 6, 7, 10

No. 1 2 3 4 5 6 7 8

35. Phenylpolyenes (continued)

Odd Systems

i = 0, 1, 2, 3, 4, 5, 6, 7, 10
No. 9 10 11 12 13 14 15 16 17

36. Phenylaryls, Biaryls, and Polyphenyls   36001-36028

1

2

3

4

5

6

7

8

9

10

11

12

13

14

15

16

36. Phenylaryls, Biaryls, and Polyphenyls (continued)

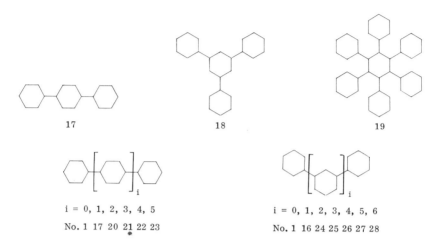

17

18

19

i = 0, 1, 2, 3, 4, 5

No. 1 17 20 21 22 23

i = 0, 1, 2, 3, 4, 5, 6

No. 1 16 24 25 26 27 28

37. Arylphenylmethyl Ions

1

2

3

4

5

6

7

8

9

10

11

37. Arylphenylmethyl Ions (continued)

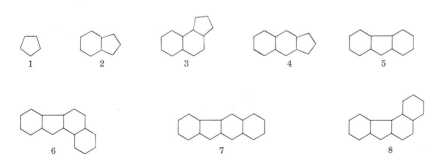

12   13   14   15

16

38. α,ω-Diphenylpolyenes   ● 38001 – 38014

Even Systems

$$\bigcirc\!\!-\!(\mathrm{C-C})_i\!-\!\bigcirc$$

i = 0, 1, 2, 3, 4, 5, 6, 7, 10,
No. 1   2   3   4   5   6   7   8   9

Odd Systems

$$\bigcirc\!\!-\!(\mathrm{C-C})_i\!-\!\mathrm{C}\!-\!\bigcirc$$

i = 0, 1, 2, 3, 4
No. 10 11 12 13 14

41. Benzocyclopentadienyl Ions   41001 – 41024

1   2   3   4   5

6   7   8

41. Benzocyclopentadienyl Ions (continued)

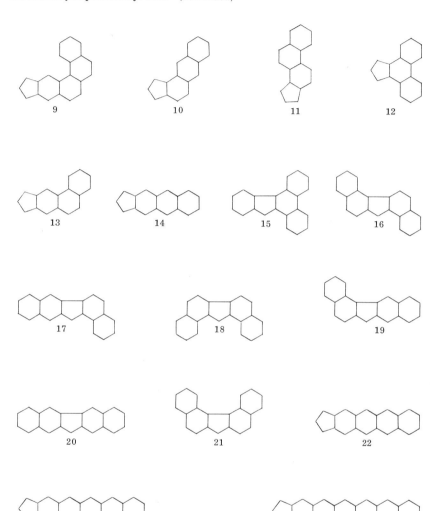

9 10 11 12

13 14 15 16

17 18 19

20 21 22

23 24

42. Benzofulvenes

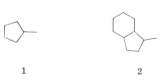 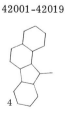

1 2 3 4

42. Benzofulvenes (continued)

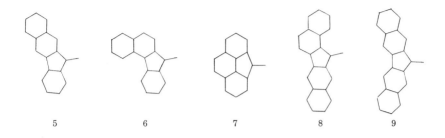

43001-43024

43. Odd Cyclopentadienylpolyenes and Their
Benzo Derivatives

43. Odd Cyclopentadienylpolyenes and Their Benzo Derivatives (continued)

$-C(-C-C)_n C$

n = 0, 1, 2, 3

No. 9 10 11 12

$-C(-C-C)_n C$

n = 0, 1, 2

No. 13 14 15

$-C(-C-C)_n C$

n = 0, 1, 2

No. 16 17 18

$-C(-C-C)_n C$

n = 0, 1, 2

No. 19 20 21

$-C(-C-C)_n C$

n = 0, 1, 2

No. 22 23 24

44. Odd α,ω-Dicyclopentadienylpolyenes and Their Benzo Derivatives

44001–44009

$-C(-C-C)_n$

n = 0, 1, 2

No. 1 2 3

$-C(-C-C)_n$

n = 0, 1, 2

No. 4 5 6

$-C(-C-C)_n$

n = 0, 1, 2

No. 7 8 9

51. Benzotropyls

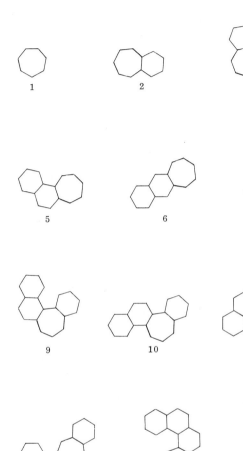

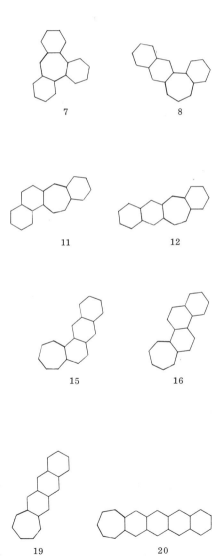

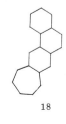
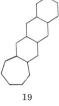
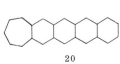

51. Benzotropyls (continued)

21   22

52. Benzoheptafulvenes

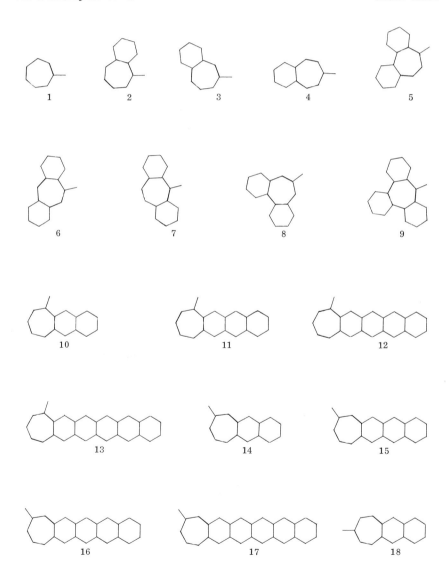

1   2   3   4   5

6   7   8   9

10   11   12

13   14   15

16   17   18

52。 Benzoheptafulvenes (continued)

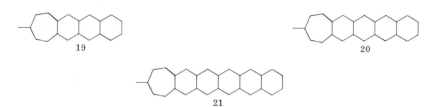

53. Polyenylcycloheptatrienyl and Vinylbenzotropyl Ions   53001-53007

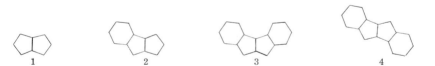

61。 Benzopentalenes   61001-61004

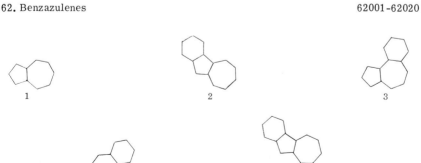

62. Benzazulenes   62001-62020

62. Benzazulenes (continued)

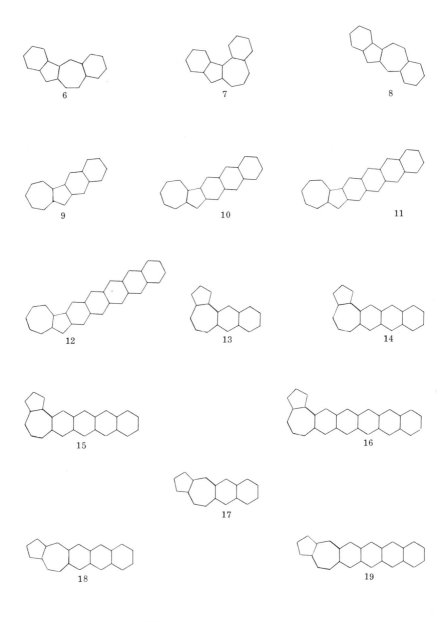

63. Benzoheptalenes

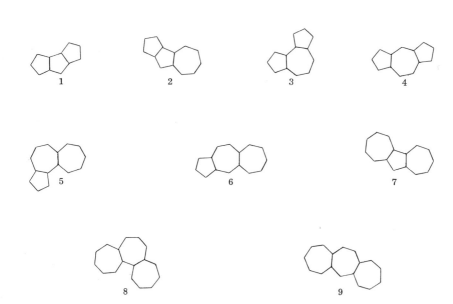

71. Cata-condensed Tricyclic Compounds

72. Indacene-like Compounds

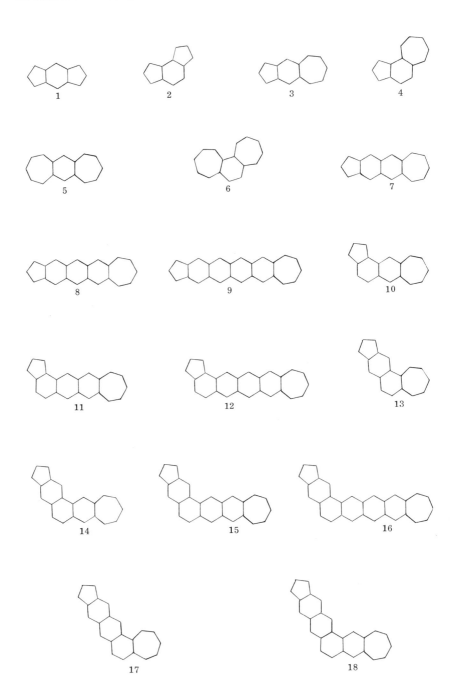

72. Indacene-like Compounds (continued)

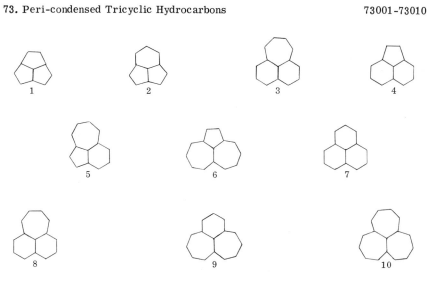

19

20

21

22

23

73. Peri-condensed Tricyclic Hydrocarbons   73001-73010

1

2

3

4

5

6

7

8

9

10

74. Fluoranthene-like Compounds   74001-74040

1

2

3

74. Flouranthene-like Compounds (continued)

4

5

6

7

8

9

10

11

12

13

14

15

16

17

18

74. Flouranthene-like Compounds   (continued)

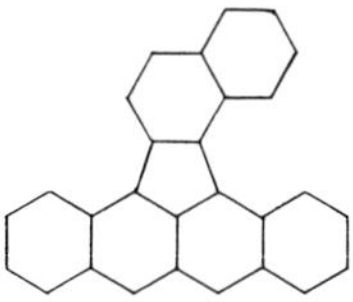

19

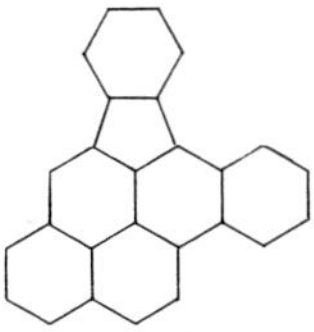

20

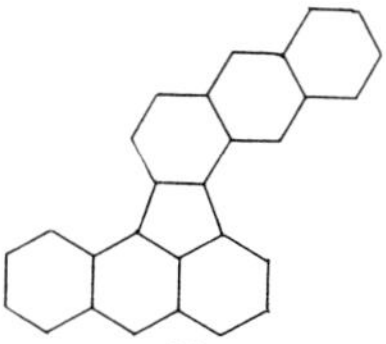

21

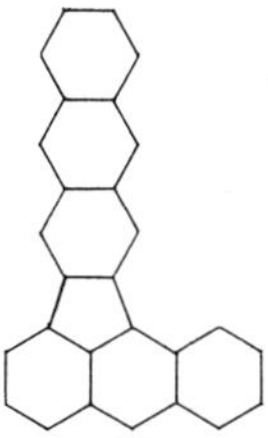

22

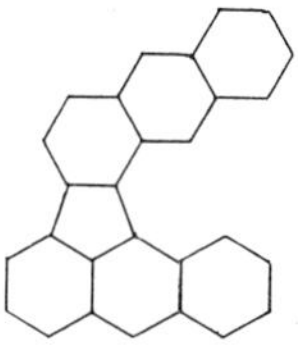

23

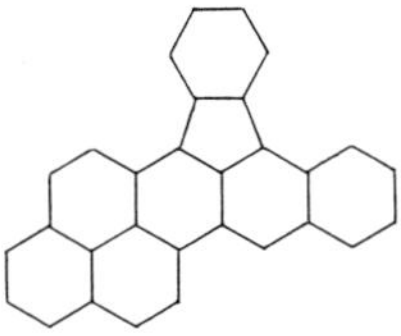

24

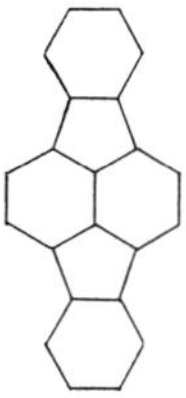

25

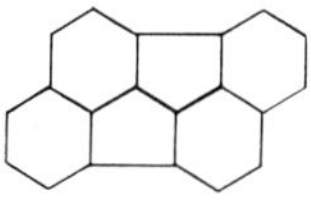

26

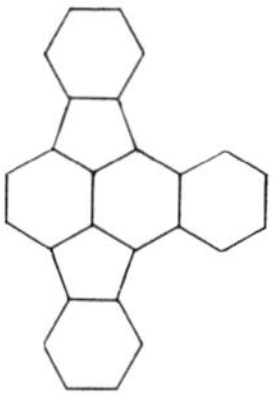

27

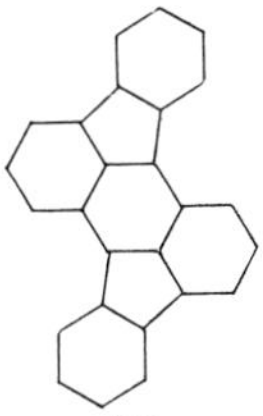

28

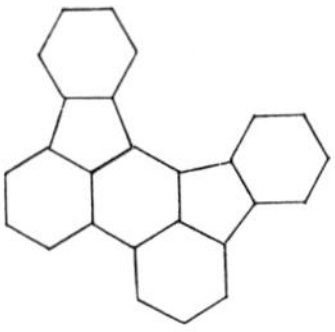

29

30

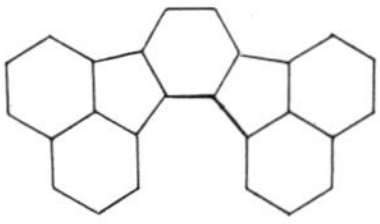

31

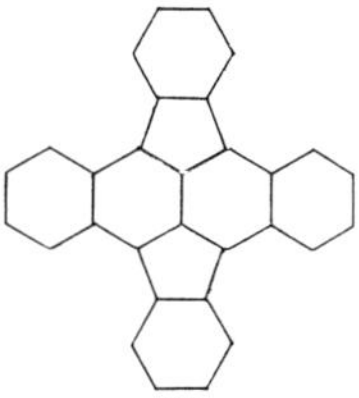

32

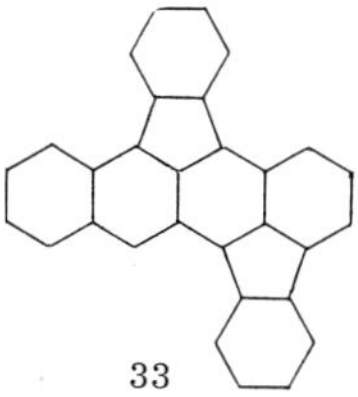

33

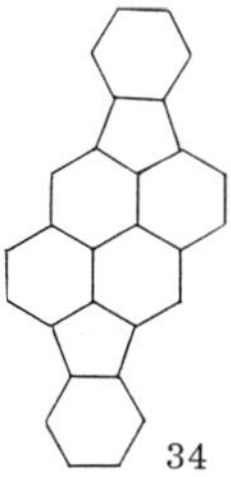

34

74. Flouranthene-like Compounds (continued)

35

36

37

38

39

40

75. Benzacenaphthylenes

1

2

3

4

5

6

7

75. Benzacenaphthylenes (continued)

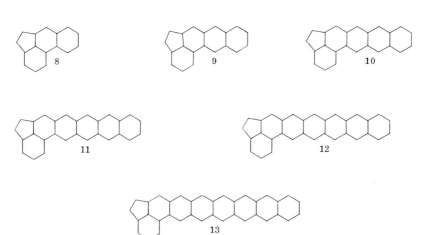

76. Fluoranthenopolyacenes   76001-76021

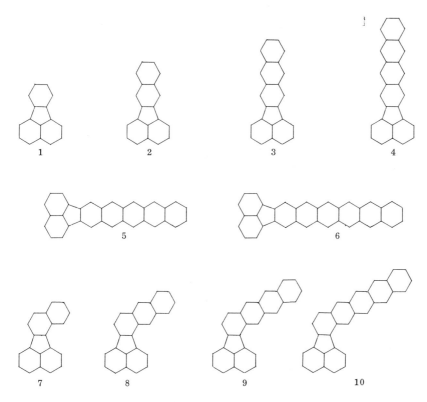

76. Fluoranthenopolyacenes (continued)

11

12

13

14

15

16

17

18

19

20

21

81. Cata-condensed Tetracyclic Hydrocarbons   81001-81017

1

2

3

81. Cata-condensed Tetracyclic Hydrocarbons (continued)

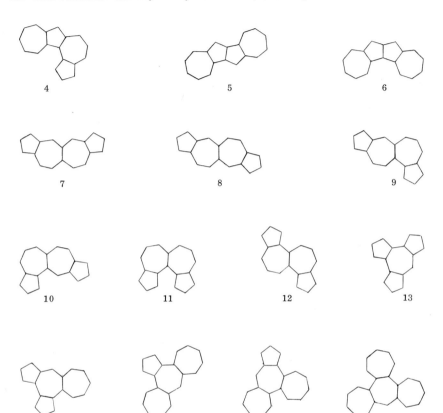

4   5   6

7   8   9

10   11   12   13

14   15   16   17

82. Peri-condensed Tetracyclic Hydrocarbons   82001-82036

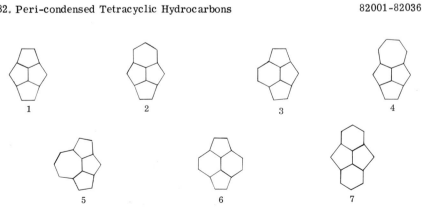

1   2   3   4

5   6   7

82. Peri-condensed Tetracyclic Hydrocarbons (continued)

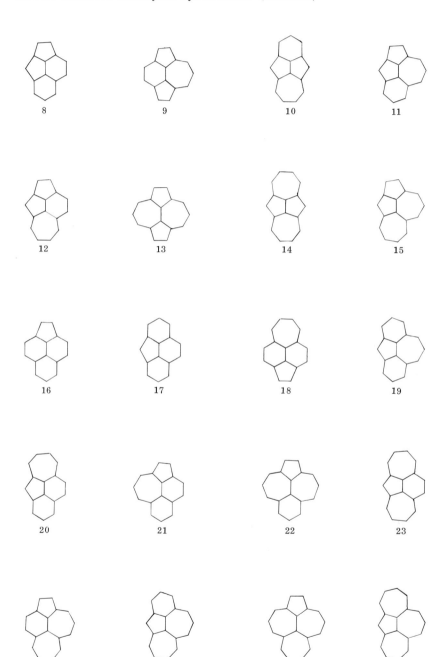

82. Peri-condensed Tetracyclic Hydrocarbons (continued)

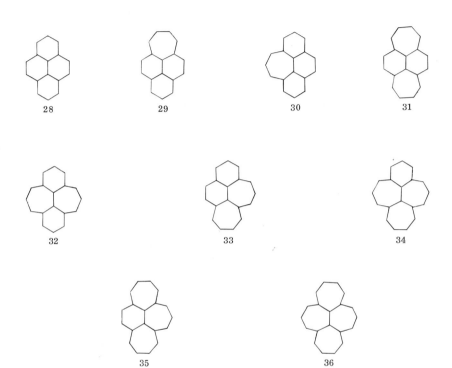

28   29   30   31

32   33   34

35   36

11. Odd Polyenes[a]

No.	Compound		n	m	W	DE_{sp}	k_2	k_1	k_{-1}	$E(N \rightarrow V_1)$
001	Allyl	C	2	2	2.8284	0.4142	–	1.4142	0.0000	1.4142
002	Pentadienyl	C	4	4	5.4641	0.3660	1.7321	1.0000	0.0000	1.0000
003	Heptatrienyl	C	6	6	8.0547	0.3424	1.4142	0.7654	0.0000	0.7654
004	Nonatetraenyl	C	8	8	10.6275	0.3284	1.1756	0.6180	0.0000	0.6180
005	Undecapentaenyl	C	10	10	13.1915	0.3192	1.0000	0.5176	0.0000	0.5176
006	Tridecahexaenyl	C	12	12	15.7505	0.3125	0.8678	0.4450	0.0000	0.4450
007	Pentadecaheptaenyl	C	14	14	18.3063	0.3076	0.7654	0.3902	0.0000	0.3902
008	Heptadecaoctaenyl	C	16	16	20.8601	0.3038	0.6840	0.3473	0.0000	0.3473
009	Nonadecanonaenyl	C	18	18	23.4124	0.3007	0.6180	0.3129	0.0000	0.3129
010	Pentacosadodecaenyl	C	24	24	31.0639	0.2943	0.4786	0.2411	0.0000	0.2411
011	Pentatriacontaheptadecaenyl	C	34	34	43.8075	0.2885	0.3473	0.1743	0.0000	0.1743

[a] The energy levels are symmetrically disposed about the k_{-1} = 0 level. In the case of anions the k_{-1} level of cation is the k_1 level of anion. Values m, W, DE_{sp} and $E(N \rightarrow V_1)$ are unchanged.

12. Even Polyenes[a]

No.	Compound	n	m	W	DE_{sp}	k_2	k_1	$E(N \rightarrow V_1)$
001	Ethylene	2	1	2.0000	0.0000	-	1.0000	2.0000
002	Butadiene	4	3	4.4721	0.1574	1.6180	0.6180	1.2361
003	Hexatriene	6	5	6.9879	0.1976	1.2470	0.4450	0.8901
004	Octatetraene	8	7	9.5175	0.2168	1.0000	0.3473	0.6946
005	Decapentaene	10	9	12.0533	0.2281	0.8308	0.2846	0.5693
006	Dodecahexaene	12	11	14.5925	0.2357	0.7092	0.2411	0.4821
007	Tetradecaheptaene	14	13	17.1335	0.2410	0.6180	0.2091	0.4181
008	Hexadecaoctaene	16	15	19.6759	0.2451	0.5473	0.1845	0.3691
009	Octadecanonaene	18	17	22.2191	0.2482	0.4910	0.1652	0.3303
010	Eicosadecaene	20	19	24.7630	0.2507	0.4450	0.1495	0.2989
011	Tetracosadodecaene	24	23	29.8519	0.2544	0.3748	0.1256	0.2512
012	Hexacosatridecaene	26	25	32.3969	0.2559	0.3473	0.1163	0.2326
013	Tetratriacontaheptadecaene	34	33	42.5784	0.2600	0.2685	0.0897	0.1795
014	Hexatriacontaoctadecaene	36	35	45.1240	0.2607	0.2540	0.0849	0.1698

[a] The energy levels are symmetrically disposed about the $k = 0$ value.

13. Odd Cyclopolyenes

No.	Compound		n	m	W	DE$_{sp}$	k$_2$	k$_1$	k$_{-1}$	k$_{-2}$	E(N→V$_1$)	Type[a]
001	Cyclopropenyl	C	2		4.0000	0.6667	-	2.0000	-1.0000	-1.0000	3.0000	H
		A	4	3	2.0000	0.0000	2.0000	-1.0000	-1.0000	-	0.0000	AH
002	Cyclopentadienyl	C	4		5.2361	0.2472	2.0000	0.6180	0.6180	-1.6180	0.0000	AH
		A	6	5	6.4721	0.4944	0.6180	0.6180	-1.6180	-1.6180	2.2361	H
003	Tropyl (Cycloheptatrienyl)	C	6		8.9879	0.4268	1.2470	1.2470	-0.4450	-0.4450	1.6920	H
		A	8	7	8.0978	0.2997	1.2470	-0.4450	-0.4450	-1.8019	0.0000	AH
004	Cyclononatetraenyl	C	8		10.8229	0.3137	1.5321	0.3473	0.3473	-1.0000	0.0000	AH
		A	10	9	11.5175	0.3908	0.3473	0.3473	-1.0000	-1.0000	1.3473	H
005	Cycloundecapentaenyl	C	10		14.0533	0.3685	0.8308	0.8308	-0.2846	-0.2846	1.1155	H
		A	12	11	13.4841	0.3167	0.8308	-0.2846	-0.2846	-1.3097	0.0000	AH
006	Cyclotridecahexaenyl	C	12		16.1103	0.3162	1.1361	0.2411	0.2411	-0.7092	0.0000	AH
		A	14	13	16.5925	0.3533	0.2411	0.2411	-0.7092	-0.7092	0.9503	H
007	Cyclopentadecaheptaenyl	C	14		19.1335	0.3422	0.6180	0.6180	-0.2091	-0.2091	0.8271	H
		A	16	15	18.7154	0.3144	0.6180	-0.2091	-0.2091	-1.0000	0.0000	AH
008	Cycloheptadecaoctaenyl	C	16		21.3068	0.3122	0.8915	0.1845	0.1845	-0.5473	0.0000	AH
		A	18	17	21.6759	0.3339	0.1845	0.1845	-0.5473	-0.5473	0.7319	H
009	Cyclononadecanonaenyl	C	18		24.2191	0.3273	0.4910	0.4910	-0.1652	-0.1652	0.6561	H
		A	20	19	23.8888	0.3099	0.4910	-0.1652	-0.1652	-0.8034	0.0000	AH
010	Cyclopentacosadodecaenyl[b]	A	26	25	31.8519	0.3141	0.1256	0.1256	-0.3748	-0.3748	0.5003	H

13. Odd Cyclopolyenes (cont.)

No.	Compound	n	m	W	DE_{sp}	k_2	k_1	k_{-1}	k_{-2}	$E(N \rightarrow V_1)$	Type[a]
011	Cyclopentatriacontahepta-decaenyl[b]	C	34	35	44.5784	0.3022	0.2685	0.2685	-0.0897	-0.0897	0.3582	H

[a] H stands for a Hückel type and AH for a anti-Hückel type compound. In a Hückel type compounds all bonding levels are double occupied and all anti-bonding levels unoccupied.

[b] These are the values for the Hückel type compounds (useful for extrapolation).

14. Even Cyclopolyenes ([n] annulenes)[a]

No.	Compound		n	m	W	DE_{sp}	k_2	k_1	$E(N \rightarrow V_1)$
001	Cyclobutadiene		4	4	4.0000	0.0000	2.0000	0.0000	0.0000
		diC	2	4	4.0000	0.5000	-	2.0000	2.0000
002	Benzene		6	6	8.0000	0.3333	1.0000	1.0000	2.0000
003	Cyclooctatetraene		8	8	9.6569	0.2071	1.4142	0.0000	0.0000
		diC	6	8	9.6569	0.4571	1.4142	1.4142	1.4142
004	Cyclodecapentaene		10	10	12.9443	0.2944	0.6180	0.6180	1.2361
005	Cyclododecahexaene		12	12	14.9282	0.2440	1.0000	0.0000	0.0000
		diC	10	12	14.9282	0.4107	1.0000	1.0000	1.0000
006	Cyclotetradecaheptaene		14	14	17.9758	0.2840	0.4450	0.4450	0.8901
007	Cyclohexadecaoctaene		16	16	20.1094	0.2568	0.7654	0.0000	0.0000
		diC	14	16	20.1094	0.3818	0.7654	0.7654	0.7654
008	Cyclooctadecanonaene		18	18	23.0351	0.2797	0.3473	0.3473	0.6946
009	Cycloeicosadecaene		20	20	25.2550	0.2628	0.6180	0.0000	0.0000
		diC	18	20	25.2550	0.3628	0.6180	0.6180	0.6180
010	Cyclohexacosatridecaene		26	26	33.1849	0.2763	0.2411	0.2411	0.4821
011	Cyclohexatriacontaocta- decaene		36	36	45.7202	0.2700	0.3473	0.0000	0.0000
		diC	34	36	45.7202	0.3256	0.3473	0.3473	0.3473

[a] The data for the corresponding dications are given in the case of anti-Hückel cyclopolyenes

(Continued on page 44)

(4n electrons), too; the values for dianions can be found easily from the data for uncharged hydrocarbons. The energy levels are symmetrically disposed about the $k = 0$ value. The energy levels $k_1 = 0$ are double degenerated.

15. Fulvenes

No.	Compound	n	m	W	DE_{sp}	k_2	k_1	k_{-1}	k_{-2}	$E(N \rightarrow V_1)$
001	Methylenecyclopropene	4	4	4.9624	0.2406	2.1701	0.3111	-1.0000	-1.4812	1.3111
002	Fulvene	6	6	7.4659	0.2443	1.0000	0.6180	-0.2541	-1.6180	0.8721
003	Heptafulvene	8	8	9.9944	0.2493	1.2470	0.2163	-0.4450	-0.7764	0.6613
004	Methylenecyclononatetraene	10	10	12.5305	0.2530	0.6430	0.3473	-0.1889	-1.0000	0.5362
005	Methylenecycloundecapentaene	12	12	15.0701	0.2558	0.8308	0.1679	-0.2846	-0.5529	0.4526

16. Fulvalenes[a]

No.	Compound		n	m	W	DE_{sp}	k_2	k_1	k_{-1}	k_{-2}	$E(N \rightarrow V_1)$
001	Trifulvalene		6	7	7.4641	0.2092	1.7321	-0.4142	-1.0000	-1.0000	0.5858
		diC	4	7	8.2925	0.6132	2.4142	1.7321	-0.4142	-1.0000	2.1463
002	Tripentafulvalene		8	9	10.9389	0.3265	0.6765	0.6180	-0.8713	-1.0000	1.4894
003	Triheptafulvalene		10	11	12.8495	0.2590	1.2125	-0.2455	-0.4450	-1.0000	0.1996
		diC	8	11	13.3405	0.4855	1.2470	1.2125	-0.2455	-0.4450	1.4580
004	Trinonafulvalene		12	13	15.9717	0.3055	0.4060	0.3473	-0.6774	-1.0000	1.0247
005	Triundecafulvalene		14	15	18.0862	0.2724	0.8308	-0.1707	-0.2846	-0.9096	0.1139
		diC	12	15	18.4276	0.4285	0.8441	0.8308	-0.1707	-0.2846	1.0015
006	Fulvalene		10	11	12.7993	0.2545	0.6180	0.6180	0.2541	-1.3028	0.3639
		diA	12	11	13.3075	0.4825	0.6180	0.2541	-1.3028	-1.6180	1.5569
007	Sesquifulvalene		12	13	15.9306	0.3024	0.6180	0.5702	-0.4450	-0.4731	1.0152
008	Pentanonafulvalene		14	15	18.0216	0.2681	0.6180	0.3473	0.1793	-0.9417	0.1680
		diA	16	15	18.3802	0.4253	0.3473	0.1793	-0.9417	-1.0000	1.1209
009	Pentaundecafulvalene		16	17	20.9909	0.2936	0.6180	0.4881	-0.2846	-0.3215	0.7728
010	Heptafulvalene		14	15	18.0047	0.2670	1.0000	-0.1826	-0.4450	-0.4450	0.2624
		diC	12	15	18.3700	0.4247	1.2470	1.0000	-0.1826	-0.4450	1.1826
011	Heptanonafulvalene		16	17	20.9769	0.2928	0.3638	0.3473	-0.4204	-0.4450	0.7677
012	Heptaundecafulvalene		18	19	23.1694	0.2721	0.7662	-0.1408	-0.2846	-0.4450	1.1438
		diC	16	19	23.4510	0.3922	0.8308	0.7662	-0.1408	-0.2846	0.9070

16. Fulvalenes[a] (cont.)

No.	Compound	n	m	W	DE$_{sp}$	k$_2$	k$_1$	k$_{-1}$	k$_{-2}$	E(N→V$_1$)
013	Nonafulvalene	18	19	23.1616	0.2717	0.3473	0.3473	0.1424	-0.8019	0.2049
	diA	20	19	23.4465	0.3919	0.3473	0.1424	-0.8019	-1.0000	0.9444
014	Nonaundecafulvalene	20	21	26.0427	0.2877	0.3473	0.3323	-0.2846	-0.2955	0.6169
015	Undecafulvalene	22	23	28.2957	0.2737	0.6665	-0.1167	-0.2846	-0.2846	0.1679
	diC	20	23	28.5291	0.3708	0.8308	0.6665	-0.1167	-0.2846	0.7832

[a] Chemical names are available only for compounds Nos. 006, 007 and 010; for other Systems "working names" were formed using the term fulvalene.

17. Radialenes (polykis-methylenecycloalkanes)

No.	Compound	n	m	W	DE_{sp}	k_2	k_1	k_{-1}	k_{-2}	$E(N \rightarrow V_1)$
001	Tris-methylenecyclopropane	6	6	7.3006	0.2168	0.6180	0.6180	-0.4142	-1.6180	1.0322
002	Tetrakis-methylenecyclo-butane	8	8	9.6569	0.2071	1.0000	0.4142	-0.4142	-1.0000	0.8284
003	Pentakis-methylenecyclo-pentane	10	10	12.1602	0.2160	0.4773	0.4773	-0.4142	-0.7376	0.8915
004	Hexakis-methylenecyclo-hexane	12	12	14.6011	0.2168	0.6180	0.4142	-0.4142	-0.6180	0.8284
005	Heptakis-methylenecyclo-heptane	14	14	17.0241	0.2160	0.4450	0.4450	-0.4142	-0.5550	0.8593
006	Octakis-methylenecyclo-octane	16	16	19.4548	0.2159	0.5176	0.4142	-0.4142	-0.5176	0.8284
007	Nonakis-methylenecyclo-nonane	18	18	21.8881	0.2160	0.4325	0.4325	-0.4142	-0.4936	0.8468
008	Decakis-methylenecyclo-decane	20	20	24.3203	0.2160	0.4773	0.4142	-0.4142	-0.4773	0.8284
009	Undecakis-methylenecyclo-undecane	22	22	26.7521	0.2160	0.4264	0.4264	-0.4142	-0.4655	0.8406

21. Benzocyclobutadienes[a]

No.	Compound	n	m	W	DE_{sp}	k_2	k_1	$E(N \longrightarrow V_1)$
001	Cyclobutadiene	4	4	4.0000	0.0000	2.0000	0.0000	0.0000
002	Bicyclo(4.2.0)octatetraene	8	9	10.3812	0.2646	1.0953	0.2624	0.5247
003	Biphenylene	12	14	16.5055	0.3218	0.8794	0.4450	0.8901
004	Benzo(a)biphenylene	16	19	22.1661	0.3245	0.8293	0.3207	0.6413
005	Benzo(b)biphenylene	16	19	22.2515	0.3290	0.6180	0.5021	1.0043
006	Dibenzo(a,g)biphenylene	20	24	27.8164	0.3257	0.7178	0.2310	0.4619
007	Dibenzo(a,i)biphenylene	20	24	27.8196	0.3258	0.7760	0.2282	0.4565
008	Dibenzo(b,h)biphenylene	20	24	27.9798	0.3325	0.5550	0.5157	1.0314

[a] The energy levels are symmetrically disposed about the $k = 0$ value.

31. Benzenoid Hydrocarbons[a]
Even Systems

No.	Compound	n	m	W	DE_{sp}	k_2'	k_1	$E(N \rightarrow V_1)$
001	Benzene	6	6	8.0000	0.3333	1.0000	1.0000	2.0000
002	Naphthalene	10	11	13.6832	0.3348	1.0000	0.6180	1.2361
003	Anthracene	14	16	19.3137	0.3321	1.0000	0.4142	0.8284
004	Phenanthrene	14	16	19.4483	0.3405	0.7691	0.6052	1.2105
005	Naphthacene	18	21	24.9308	0.3300	0.7775	0.2950	0.5899
006	Benz(a)anthracene	18	21	25.1012	0.3382	0.7150	0.4523	0.9046
007	Chrysene	18	21	25.1922	0.3425	0.7923	0.5201	1.0403
008	Benzo(c)phenanthrene	18	21	25.1875	0.3423	0.6622	0.5676	1.1352
009	Triphenylene	18	21	25.2745	0.3464	0.6840	0.6840	1.3681
010	Pyrene	16	19	22.5055	0.3424	0.8794	0.4450	0.8901
011	Pentacene	22	26	30.5440	0.3286	0.6180	0.2197	0.4394
012	Benzo(a)naphthacene	22	26	30.7256	0.3356	0.6874	0.3271	0.6541
013	Pentaphene	22	26	30.7627	0.3370	0.5209	0.4372	0.8743
014	Benzo(b)chrysene	22	26	30.8390	0.3400	0.7045	0.4048	0.8096
015	Dibenzo(b,g)phenanthrene	22	26	30.8338	0.3398	0.6601	0.4186	0.8372
016	Dibenz(a,c)anthracene	22	26	30.9418	0.3439	0.7140	0.4991	0.9982
017	Dibenz(a,j)anthracene	22	26	30.8795	0.3415	0.6180	0.4917	0.9835
018	Dibenz(a,h)anthracene	22	26	30.8805	0.3416	0.6843	0.4735	0.9470
019	Picene	22	26	30.9432	0.3440	0.6803	0.5019	1.0038
020	Benzo(g)chrysene	22	26	30.9990	0.3461	0.7107	0.5319	1.0638

31. Benzenoid Hydrocarbons[a] (cont. 1)

No.	Compound	n	m	W	DE_{sp}	k_2	k_1	$E(N \rightarrow V_1)$
021	Benzo(c)chrysene	22	26	30.9386	0.3438	0.6026	0.5498	1.0997
022	Dibenzo(c,g)phenanthrene	22	26	30.9362	0.3437	0.6567	0.5354	1.0709
023	Benzo(e)pyrene	20	24	28.2220	0.3426	0.8018	0.3711	0.7423
024	Benzo(e)pyrene	20	24	28.3361	0.3473	0.7181	0.4970	0.9939
025	Perylene	20	24	28.2453	0.3436	1.0000	0.3473	0.6946
026	Hexacene	26	31	36.1560	0.3276	0.5015	0.1694	0.3387
027	Benzo(a)pentacene	26	31	36.3413	0.3336	0.6176	0.2436	0.4872
028	Dibenzo(a,c)naphthacene	26	31	36.5704	0.3410	0.7265	0.3557	0.7114
029	Dibenzo(a,j)naphthacene	26	31	36.5172	0.3393	0.6800	0.3584	0.7168
030	Dibenzo(a,l)naphthacene	26	31	36.5169	0.3393	0.6323	0.3607	0.7213
031	Hexaphene	26	31	36.3905	0.3352	0.4856	0.3357	0.6715
032	Naphtho(2,1-a)naphthacene	26	31	36.4613	0.3375	0.6226	0.3033	0.6066
033	Dibenzo(b,k)chrysene	26	31	36.4839	0.3382	0.5702	0.3482	0.6965
034	Benzo(h)pentaphene	26	31	36.6142	0.3424	0.5207	0.5058	1.0115
035	Tribenz(a,c,h)anthracene	26	31	36.7149	0.3456	0.6371	0.5224	1.0449
036	Dibenzo(g,p)chrysene	26	31	36.7953	0.3482	0.7046	0.5115	1.0229
037	Benzo(c)picene	26	31	36.6920	0.3449	0.6954	0.4714	0.9428
038	Benzo(a)perylene	24	29	33.9165	0.3419	0.7771	0.2648	0.5296
039	Dibenzo(de,mn)naphthacene	24	29	33.7982	0.3379	0.7755	0.1987	0.3974
040	Naphtho(8,1,2-cde)naphthacene	24	29	33.8634	0.3401	0.6476	0.3026	0.6052
041	Dibenzo(de,qr)naphthacene	24	29	34.0059	0.3450	0.5133	0.5053	1.0106
042	Dibenzo(fg,op)naphthacene	24	29	34.1644	0.3505	0.6733	0.5550	1.1099

31. Benzenoid Hydrocarbons[a] (cont. 2)

No.	Compound	n	m	W	DE_{sp}	k_2	k_1	$E(N \rightarrow V_1)$
043	Benzo(b)perylene	24	29	34.0070	0.3451	0.7251	0.3514	0.7028
044	Naphtho(1,2,3,4-def)chrysene	24	29	34.0646	0.3471	0.7424	0.4216	0.8432
045	Benzo(ghi)perylene	22	27	31.4251	0.3491	0.6843	0.4392	0.8784
046	Dibenzo(b,def)chrysene	24	29	33.9278	0.3423	0.7932	0.3027	0.6053
047	Benzo(rst)pentaphene	24	29	33.9542	0.3432	0.6818	0.3420	0.6841
048	Dibenzo(def,p)chrysene	24	29	34.0307	0.3459	0.6690	0.3983	0.7966
049	Dibenzo(def,mno)chrysene	22	27	31.2529	0.3427	0.7500	0.2910	0.5819
050	Heptacene	30	36	41.7675	0.3269	0.4142	0.1342	0.2684
051	Benzo(a)hexacene	30	36	41.9542	0.3321	0.5195	0.1871	0.3741
052	Benzo(p)hexaphene	30	36	42.0076	0.3335	0.4790	0.2510	0.5021
053	Trinaphthylene	30	36	42.2903	0.3414	0.5279	0.5157	1.0314
054	Dibenzo(a,c)pentacene	30	36	42.1875	0.3385	0.6597	0.2621	0.5242
055	Dibenzo(a,n)pentacene	30	36	42.1373	0.3371	0.6052	0.2695	0.5390
056	Dibenzo(a,l)pentacene	30	36	42.1374	0.3371	0.6278	0.2690	0.5381
057	Naphtho(8,1,2-cde)pentacene	28	34	39.4847	0.3378	0.5296	0.2426	0.4852
058	Dibenzo(de,uv)pentacene	28	34	39.6357	0.3422	0.5190	0.3587	0.7174
059	Dibenzo(fg,st)pentacene	28	34	39.8331	0.3480	0.5738	0.5009	1.0019
060	Heptaphene	30	36	42.0201	0.3339	0.3617	0.3256	0.6512
061	Tetrabenz(a,c,h,j)anthracene	30	36	42.5461	0.3485	0.6052	0.5774	1.1548
062	Benzo(a)naphtho(8,1,2-cde)-naphthacene	28	34	39.7089	0.3444	0.6580	0.3414	0.6828
063	Dibenzo(b,n)perylene	28	34	39.7689	0.3461	0.7172	0.3553	0.7106

31. Benzenoid Hydrocarbons[a] (cont. 3)

No.	Compound	n	m	W	DE$_{sp}$	k$_2$	k$_1$	E(N→V$_1$)
064	Dibenzo(a,f)perylene	28	34	39.5809	0.3406	0.7294	0.1934	0.3868
065	Dibenzo(a,k)perylene	28	34	39.6773	0.3434	0.6902	0.2684	0.5368
066	Dibenzo(a,e)perylene	28	34	39.6950	0.3440	0.6731	0.2858	0.5716
067	Dibenzo(b,e)perylene	28	34	39.7415	0.3453	0.5586	0.3658	0.7316
068	Benzo(vwx)hexaphene	28	34	39.6014	0.3412	0.5614	0.2969	0.5937
069	Benzo(a)naphtho(2,1,8-hij)-naphthacene	28	34	39.6492	0.3426	0.6769	0.3186	0.6371
070	Tribenzo(a,c,j)naphthacene	30	36	42.3598	0.3433	0.6557	0.3900	0.7800
071	Dibenzo(de,op)pentacene	28	34	39.3841	0.3348	0.5868	0.1280	0.2560
072	Dibenzo(a,n)perylene	28	34	39.6778	0.3435	0.7243	0.2673	0.5347
073	Dibenzo(a,j)perylene	28	34	39.5936	0.3410	0.6377	0.2135	0.4269
074	Dibenzo(a,o)perylene	28	34	39.5988	0.3411	0.7042	0.2089	0.4179
075	Dibenzo(b,k)perylene	28	34	39.7688	0.3461	0.6964	0.3557	0.7114
076	Tribenzo(b,e,h)pyrene	28	34	39.7382	0.3452	0.5468	0.3730	0.7460
077	Dibenzo(cd,lm)perylene	26	32	37.0896	0.3466	0.8280	0.2846	0.5693
078	Coronene	24	30	34.5718	0.3524	0.5392	0.5392	1.0784
079	Octacene	34	41	47.3789	0.3263	0.3473	0.1088	0.2176
080	Octaphene	34	41	47.6380	0.3326	0.3458	0.2534	0.5068
081	Benzo(j)heptaphene	34	41	47.8765	0.3385	0.3730	0.3587	0.7175
082	Tribenzo(cd,ghi,lm)perylene	28	35	40.1048	0.3459	0.6995	0.2229	0.4457
083	Phenanthro(1,10,9,8-fghij)-perylene	28	35	40.0779	0.3451	0.7046	0.1774	0.3549

31. Benzenoid Hydrocarbons[a] (cont. 4)

No.	Compound	n	m	W	DE_{sp}	k_2	k_1	$E(N \rightarrow V_1)$
084	Tribenzo(a,c,l)pentacene	34	41	47.9827	0.3410	0.6410	0.2898	0.5795
085	Tetrabenzo(a,c,j,l)naphthacene	34	41	48.2014	0.3464	0.6548	0.4247	0.8495
086	Naphtho(8,1,2-cde)hexacene	32	39	45.0995	0.3359	0.4502	0.1936	0.3872
087	Dibenzo(de,yz)hexacene	32	39	45.2534	0.3398	0.5211	0.2641	0.5283
088	Dibenzo(fg,wx)hexacene	32	39	45.4624	0.3452	0.5806	0.3570	0.7140
089	Dibenzo(hi,uv)hexacene	32	39	45.5014	0.3462	0.5356	0.4762	0.9523
090	Tetrabenzo(a,de,j,mn)naphthacene	32	39	45.2105	0.3387	0.6517	0.1252	0.2505
091	Pyranthrene	30	37	42.7407	0.3443	0.6763	0.2580	0.5159
092	Tribenzo(a,ghi,o)perylene	30	37	42.8188	0.3465	0.6923	0.2941	0.5883
093	Benzo(xyz)heptaphene	32	39	45.2513	0.3398	0.4722	0.2734	0.5469
094	Tribenzo(de,kl,rst)pentaphene	30	37	42.8389	0.3470	0.7092	0.2411	0.4821
095	Nonaphene	38	46	53.2565	0.3317	0.2640	0.2477	0.4955
096	Tetrabenzo(a,de,l,op)pentacene	36	44	50.8226	0.3369	0.5365	0.0833	0.1666
097	Tetrabenzo(a,c,l,n)pentacene	38	46	53.8277	0.3441	0.6840	0.3126	0.6252
098	Dibenzo(j,xyz)heptaphene	36	44	51.1041	0.3433	0.4641	0.3180	0.6360
099	Dibenzo(hi,yz)heptacene	36	44	51.1306	0.3439	0.5076	0.3549	0.7098
100	Dinaphtho(1,2,3-cd:1',2',3'-lm)-perylene	34	42	48.5320	0.3460	0.6311	0.2265	0.4530
101	Dinaphtho(1,2,3-cd:3',2',1'-lm)-perylene	34	42	48.5384	0.3462	0.5984	0.2353	0.4706
102	Tetrabenzo(a,f,j,o)perylene	36	44	50.9590	0.3400	0.6324	0.1199	0.2397
103	Dibenzo(bc,kl)coronene	30	38	43.1197	0.3453	0.5389	0.1859	0.3718

31. Benzenoid Hydrocarbons[a] (cont. 5)

No.	Compound	n	m	W	DE_{sp}	k_2	k_1	$E(N \rightarrow V_1)$
104	Dibenzo(bc,ef)coronene	30	38	43.3000	0.3500	0.7119	0.2539	0.5078
105	Dibenzo(jk,a'b')octacene	40	49	56.7597	0.3420	0.3766	0.3391	0.6781
106	Ovalene	32	41	46.4974	0.3536	0.6052	0.3359	0.6718

[a] The energy levels are symmetrically disposed about the k = 0 value.

31. Benzenoid Hydrocarbons

Odd Systems

No.	Compound		n	m	W	DE_{sp}	k_2	k_1	k_{-1}	k_{-2}	$E(N \rightarrow V_1)$
107	Phenalenyl	C	12	15	17.8272	0.3885	1.0000	1.0000	0.0000	-1.0000	1.0000
108	Benz(de)anthryl	C	16	20	23.5818	0.3791	1.0000	0.7182	0.0000	-0.7182	0.7182
109	Dibenzo(def,jk)phenanthryl	C	18	23	26.6964	0.3781	0.8842	0.6622	0.0000	-0.6622	0.6622
110	Dibenzo(b,e)phenalenyl	C	20	25	29.3027	0.3721	1.0000	0.5576	0.0000	-0.5576	0.5576
111	Benzo(fg)naphthacenyl	C	20	25	29.3406	0.3736	0.7261	0.7251	0.0000	-0.7251	0.7251
112	Dibenz(a,de)anthryl	C	20	25	29.3368	0.3735	0.8075	0.6548	0.0000	-0.6548	0.6548
113	Tribenzo(a,e,i)phenalenyl	C	24	30	35.0961	0.3699	0.6741	0.6741	0.0000	-0.6741	0.6741
114	Dibenzo(a,fg)naphthacenyl	C	24	30	35.1057	0.3702	0.7517	0.6653	0.0000	-0.6653	0.6653
115	Tribenz(a,de,j)anthryl	C	24	30	35.0967	0.3699	0.7920	0.5981	0.0000	-0.5981	0.5981

31107-31115

32. Arylmethyl Ions

No.	Compound	n	m	W	DE_sp	k_2	k_1	k_{-1}	k_{-2}	E(N→V$_1$)
001	Benzyl	6	7	8.7206	0.3887	1.2593	1.0000	-1.0000	-1.0000	1.0000
002	1-Naphthylmethyl	10	12	14.4954	0.3746	1.0000	0.7991	0.0000	-0.7991	0.7991
003	2-Naphthylmethyl	10	12	14.4269	0.3689	1.1232	0.6757	0.0000	-0.6757	0.6757
004	1-Anthrylmethyl	14	17	20.1618	0.3625	1.0000	0.5370	0.0000	-0.5370	0.5370
005	2-Anthrylmethyl	14	17	20.0834	0.3578	1.0000	0.4784	0.0000	-0.4784	0.4784
006	9-Anthrylmethyl	14	17	20.2648	0.3685	1.0000	0.6826	0.0000	-0.6826	0.6826
007	4-Phenanthrylmethyl	14	17	20.2312	0.3665	0.8930	0.6360	0.0000	-0.6360	0.6360
008	3-Phenanthrylmethyl	14	17	20.2032	0.3649	0.7701	0.6992	0.0000	-0.6992	0.6992
009	2-Phenanthrylmethyl	14	17	20.1847	0.3638	0.8862	0.6063	0.0000	-0.6063	0.6063
010	1-Phenanthrylmethyl	14	17	20.2515	0.3677	0.8695	0.6774	0.0000	-0.6774	0.6774
011	9-Phenanthrylmethyl	14	17	20.2616	0.3683	0.8178	0.7421	0.0000	-0.7421	0.7421
012	5-Naphthacenylmethyl	18	22	25.9380	0.3608	0.8413	0.5224	0.0000	-0.5224	0.5224
013	1-Naphthacenylmethyl	18	22	25.7976	0.3544	0.8866	0.3838	0.0000	-0.3838	0.3838
014	2-Naphthacenylmethyl	18	22	25.7192	0.3509	0.8040	0.3552	0.0000	-0.3552	0.3552
015	1-Benz(a)anthrylmethyl	18	22	25.8818	0.3583	0.8707	0.4523	0.0000	-0.4523	0.4523
016	2-Benz(a)anthrylmethyl	18	22	25.8642	0.3575	0.7515	0.4978	0.0000	-0.4978	0.4978
017	3-Benz(a)anthrylmethyl	18	22	25.8421	0.3565	0.7739	0.4617	0.0000	-0.4617	0.4617
018	4-Benz(a)anthrylmethyl	18	22	25.9060	0.3594	0.8671	0.4744	0.0000	-0.4744	0.4744
019	5-Benz(a)anthrylmethyl	18	22	25.9346	0.3607	0.8257	0.5319	0.0000	-0.5319	0.5319
020	6-Benz(a)anthrylmethyl	18	22	25.9340	0.3606	0.7688	0.5430	0.0000	-0.5430	0.5430
021	7-Benz(a)anthrylmethyl	18	22	26.0303	0.3650	0.7243	0.7006	0.0000	-0.7006	0.7006
022	8-Benz(a)anthrylmethyl	18	22	25.9409	0.3609	0.7300	0.5714	0.0000	-0.5714	0.5714
023	9-Benz(a)anthrylmethyl	18	22	25.8608	0.3573	0.7663	0.4934	0.0000	-0.4934	0.4934

32. Arylmethyl Ions (cont. 1)

No.	Compound	n	m	W	DE_{sp}	k_2	k_1	k_{-1}	k_{-2}	$E(N \rightarrow V_1)$
024	10-Benz(a)anthrylmethyl	18	22	25.8675	0.3576	0.7183	0.5202	0.0000	-0.5202	0.5202
025	11-Benz(a)anthrylmethyl	18	22	25.9345	0.3607	0.7557	0.5474	0.0000	-0.5474	0.5474
026	12-Benz(a)anthrylmethyl	18	22	26.0037	0.3638	0.8026	0.6043	0.0000	-0.6043	0.6043
027	2-Triphenylenylmethyl	18	22	26.0191	0.3645	0.7673	0.6840	0.0000	-0.6840	0.6840
028	1-Triphenylenylmethyl	18	22	26.0517	0.3660	0.7482	0.6840	0.0000	-0.6840	0.6840
029	5-Chrysenylmethyl	18	22	25.9824	0.3628	0.8261	0.5663	0.0000	-0.5663	0.5663
030	6-Chrysenylmethyl	18	22	26.0277	0.3649	0.7939	0.6832	0.0000	-0.6832	0.6832
031	1-Chrysenylmethyl	18	22	26.0030	0.3638	0.7959	0.5990	0.0000	-0.5990	0.5990
032	2-Chrysenylmethyl	18	22	25.9309	0.3605	0.8485	0.5329	0.0000	-0.5329	0.5329
033	3-Chrysenylmethyl	18	22	25.9494	0.3613	0.8097	0.5730	0.0000	-0.5730	0.5730
034	4-Chrysenylmethyl	18	22	25.9829	0.3629	0.8072	0.5710	0.0000	-0.5710	0.5710
035	1-Benzo(c)phenanthrylmethyl	18	22	25.9858	0.3630	0.6934	0.6427	0.0000	-0.6427	0.6427
036	2-Benzo(c)phenanthrylmethyl	18	22	25.9392	0.3609	0.7421	0.5834	0.0000	-0.5834	0.5834
037	3-Benzo(c)phenanthrylmethyl	18	22	25.9327	0.3606	0.6745	0.6124	0.0000	-0.6124	0.6124
038	4-Benzo(c)phenanthrylmethyl	18	22	25.9935	0.3633	0.7348	0.6153	0.0000	-0.6153	0.6153
039	5-Benzo(c)phenanthrylmethyl	18	22	26.0013	0.3637	0.8004	0.6125	0.0000	-0.6125	0.6125
040	6-Benzo(c)phenanthrylmethyl	18	22	25.9900	0.3632	0.8367	0.5730	0.0000	-0.5730	0.5730
041	4-Pyrenylmethyl	16	20	23.3319	0.3666	0.8880	0.5469	0.0000	-0.5469	0.5469
042	1-Pyrenylmethyl	16	20	23.3730	0.3686	0.9145	0.6124	0.0000	-0.6124	0.6124
043	2-Pyrenylmethyl	16	20	23.2226	0.3611	1.0000	0.4450	0.0000	-0.4450	0.4450
044	2-Biphenylmethyl	12	14	17.1505	0.3679	1.0000	0.7586	0.0000	-0.7586	0.7586
045	3-Biphenylmethyl	12	14	17.1017	0.3644	1.0000	0.7169	0.0000	-0.7169	0.7169
046	4-Biphenylmethyl	12	14	17.1392	0.3671	1.0000	0.8437	0.0000	-0.8437	0.8437

32. Arylmethyl Ions (cont. 2)

No.	Compound	n	m	W	DE_{sp}	k_2	k_1	k_{-1}	k_{-2}	$E(N \rightarrow V_1)$
047	1-Phenalenylmethyl	14	16	19.0895	0.3181	1.0000	0.3249	-0.3249	-1.0000	0.6497
048	2-Phenalenylmethyl	14	16	18.5390	0.2837	1.0000	0.0000	0.0000	-1.0000	0.0000
049	9-Benz(de)anthrylmethyl	18	21	24.3172	0.3008	0.7806	0.0000	0.0000	-0.7806	0.0000
050	8-Benz(de)anthrylmethyl	18	21	24.6789	0.3180	0.8570	0.1791	-0.1791	-0.8570	0.3582
051	7-Benz(de)anthrylmethyl	18	21	24.9607	0.3315	0.8151	0.3839	-0.3839	-0.8151	0.7679
052	6-Benz(de)anthrylmethyl	18	21	24.8502	0.3262	0.7190	0.3301	-0.3301	-0.7190	0.6602
053	5-Benz(de)anthrylmethyl	18	21	24.2967	0.2998	0.7324	0.0000	0.0000	-0.7324	0.0000
054	4-Benz(de)anthrylmethyl	18	21	24.8510	0.3262	0.7359	0.3277	-0.3277	-0.7359	0.6555
055	3-Benz(de)anthrylmethyl	18	21	24.7718	0.3225	0.8439	0.2442	-0.2442	-0.8439	0.4883
056	2-Benz(de)anthrylmethyl	18	21	24.3009	0.3000	0.7447	0.0000	0.0000	-0.7447	0.0000
057	1-Benz(de)anthrylmethyl	18	21	24.7606	0.3219	0.7513	0.2523	-0.2523	-0.7513	0.5047
058	11-Benz(de)anthrylmethyl	18	21	24.3599	0.3029	0.8689	0.0000	0.0000	-0.8689	0.0000
059	10-Benz(de)anthrylmethyl	18	21	24.6382	0.3161	0.7406	0.1922	-0.1922	-0.7406	0.3844
060	3-Naphth(3,2,1-de)anthrylmethyl	22	26	30.0460	0.3095	0.6034	0.0000	0.0000	-0.6034	0.0000
061	4-Naphth(3,2,1-de)anthrylmethyl	22	26	30.3688	0.3219	0.6489	0.1475	-0.1475	-0.6489	0.2950
062	5-Naphth(3,2,1-de)anthrylmethyl	22	26	30.6276	0.3318	0.7263	0.2886	-0.2886	-0.7263	0.5771
063	6-Naphth(3,2,1-de)anthrylmethyl	22	26	30.5841	0.3302	0.5605	0.3420	-0.3420	-0.5605	0.6841
064	7-Naphth(3,2,1-de)anthrylmethyl	22	26	30.0239	0.3086	0.5785	0.0000	0.0000	-0.5785	0.0000
065	1-Naphth(3,2,1-de)anthrylmethyl	22	26	30.0986	0.3115	0.6527	0.0000	0.0000	-0.6527	0.0000
066	2-Naphth(3,2,1-de)anthrylmethyl	22	26	30.3226	0.3201	0.5903	0.1602	-0.1602	-0.5903	0.3204
067	3-Dibenz(a,de)anthrylmethyl	22	26	30.0735	0.3105	0.6919	0.0000	0.0000	-0.6919	0.0000
068	4-Dibenz(a,de)anthrylmethyl	22	26	30.4376	0.3245	0.6984	0.1819	-0.1819	-0.6984	0.3637

32. Arylmethyl Ions (cont. 3)

No.	Compound	n	m	W	DEsp	k_2	k_1	k_{-1}	k_{-2}	E(N→V_1)
069	5-Dibenz(a,de)anthrylmethyl	22	26	30.7229	0.3355	0.7119	0.3882	-0.3882	-0.7119	0.7764
070	6-Dibenz(a,de)anthrylmethyl	22	26	30.5484	0.3288	0.6909	0.2633	-0.2633	-0.6909	0.5266
071	7-Dibenz(a,de)anthrylmethyl	22	26	30.0605	0.3100	0.6916	0.0000	0.0000	-0.6916	0.0000
072	8-Dibenz(a,de)anthrylmethyl	22	26	30.5380	0.3284	0.6592	0.2724	-0.2724	-0.6592	0.5447
073	9-Dibenz(a,de)anthrylmethyl	22	26	30.1169	0.3122	0.7116	0.0000	0.0000	-0.7116	0.0000
074	10-Dibenz(a,de)anthrylmethyl	22	26	30.3436	0.3209	0.7115	0.1510	-0.1510	-0.7115	0.3020
075	11-Dibenz(a,de)anthrylmethyl	22	26	30.0722	0.3874	0.6674	0.0000	0.0000	-0.6674	0.0000
076	12-Dibenz(a,de)anthrylmethyl	22	26	30.3854	0.3225	0.7394	0.1418	-0.1418	-0.7394	0.2835
077	13-Dibenz(a,de)anthrylmethyl	22	26	30.6269	0.3318	0.7909	0.2882	-0.2882	-0.7909	0.5764
078	1-Dibenz(a,de)anthrylmethyl	22	26	30.1157	0.3891	0.7067	0.0000	0.0000	-0.7067	0.0000
079	2-Dibenz(a,de)anthrylmethyl	22	26	30.3972	0.3230	0.6646	0.1953	-0.1953	-0.6646	0.3907
080	6-Benzo(fg)naphthacenylmethyl	22	26	30.0763	0.3876	0.7254	0.0000	0.0000	-0.7254	0.0000
081	7-Benzo(fg)naphthacenylmethyl	22	26	30.4391	0.3246	0.7257	0.1807	-0.1807	-0.7257	0.3615
082	8-Benzo(fg)naphthacenylmethyl	22	26	30.7229	0.3355	0.7258	0.3880	-0.3880	-0.7258	0.7760
083	2-Benzo(fg)naphthacenylmethyl	22	26	30.0671	0.3872	0.7261	0.0000	0.0000	-0.7261	0.0000
084	1-Benzo(fg)naphthacenylmethyl	22	26	30.4370	0.3245	0.7252	0.1812	-0.1812	-0.7252	0.3625
085	4-Benzo(fg)naphthacenylmethyl	22	26	30.1177	0.3891	0.7256	0.0000	0.0000	-0.7256	0.0000
086	5-Benzo(fg)naphthacenylmethyl	22	26	30.3992	0.3230	0.7260	0.1935	-0.1935	-0.7260	0.3870
087	4-Tribenzo(a,e,i)phenalenyl-methyl	26	31	35.8756	0.3831	0.6741	0.0000	0.0000	-0.6741	0.0000
088	3-Tribenzo(a,e,i)phenalenyl-methyl	26	31	36.1156	0.3263	0.6741	0.1609	-0.1609	-0.6741	0.3218
089	2-Tribenzo(a,e,i)phenalenyl-methyl	26	31	35.8327	0.3817	0.6741	0.0000	0.0000	-0.6741	0.0000

32. Arylmethyl Ions (cont. 4)

No.	Compound	n	m	W	DE_{sp}	k_2	k_1	k_{-1}	k_{-2}	$E(N \rightarrow V_1)$
090	1-Tribenzo(a,e,i)phenalenyl-methyl	26	31	36.1561	0.3276	0.6741	0.1512	-0.1512	-0.6741	0.3024
091	5-Tribenzo(a,e,i)phenalenyl-methyl	26	31	36.4069	0.3357	0.6741	0.3118	-0.3118	-0.6741	0.6235
092	6-Naphth(2,1,8,7-defg)anthryl-methyl	20	24	28.0680	0.3362	0.6622	0.3906	-0.3906	-0.6622	0.7812
093	5-Naphth(2,1,8,7-defg)anthryl-methyl	20	24	27.9271	0.3303	0.7421	0.2753	-0.2753	-0.7421	0.5506
094	4-Naphth(2,1,8,7-defg)anthryl-methyl	20	24	27.4127	0.3922	0.6678	0.0000	0.0000	-0.6678	0.0000
095	3-Naphth(2,1,8,7-defg)anthryl-methyl	20	24	27.9231	0.3301	0.7079	0.2801	-0.2801	-0.7079	0.5603
096	2-Naphth(2,1,8,7-defg)anthryl-methyl	20	24	27.7873	0.3245	0.7179	0.1785	-0.1785	-0.7179	0.3569
097	1-Naphth(2,1,8,7-defg)anthryl-methyl	20	24	27.4916	0.3955	0.8132	0.0000	0.0000	-0.8132	0.0000
098	6-Benzo(cd)naphtho(8,1,2-ijk)-pyrenylmethyl	26	32	36.8585	0.3393	0.5130	0.3362	-0.3362	-0.5130	0.6723
099	5-Benzo(cd)naphtho(8,1,2-ijk)-pyrenylmethyl	26	32	36.6880	0.3340	0.5661	0.2098	-0.2098	-0.5661	0.4195
100	4-Benzo(cd)naphtho(8,1,2-ijk)-pyrenylmethyl	26	32	36.2171	0.3818	0.4641	0.0000	0.0000	-0.4641	0.0000
101	3-Benzo(cd)naphtho(8,1,2-ijk)-pyrenylmethyl	26	32	36.6815	0.3338	0.5324	0.2161	-0.2161	-0.5324	0.4321
102	2-Benzo(cd)naphtho(8,1,2-ijk)-pyrenylmethyl	26	32	36.5035	0.3282	0.5182	0.1060	-0.1060	-0.5182	0.2120
103	1-Benzo(cd)naphtho(8,1,2-ijk)-pyrenylmethyl	26	32	36.3233	0.3851	0.5513	0.0000	0.0000	-0.5513	0.0000

32. Arylmethyl Ions (cont. 5)

No.	Compound	n	m	W	DE_{sp}	k_2	k_1	k_{-1}	k_{-2}	$E(N \rightarrow V_1)$
104	13-Benzo(cd)naphtho(8,1,2-ijk)-pyrenylmethyl	26	32	36.4296	0.3884	0.7174	0.0000	0.0000	-0.7174	0.0000
105	5-Tribenz(a,de,j)anthrylmethyl	26	31	36.4719	0.3378	0.7016	0.3598	-0.3598	-0.7016	0.7197
106	14-Dibenzo(a,fg)naphthacenyl-	26	31	36.4718	0.3378	0.7316	0.3595	-0.3595	-0.7316	0.7190

33. Vinylpolyacenes[a]

No.	Compound	n	m	W	DE_{sp}	k_2	k_1	$E(N \rightarrow V_1)$
001	Styrene	8	8	10.4243	0.3030	1.0000	0.6622	1.3243
002	1-Vinylnaphthalene	12	13	16.1294	0.3176	1.0000	0.4985	0.9969
003	2-Vinylnaphthalene	12	13	16.1113	0.3163	0.7922	0.5463	1.0926
004	1-Vinylanthracene	16	18	21.7643	0.3202	0.7815	0.3710	0.7421
005	2-Vinylanthracene	16	18	21.7448	0.3192	0.7451	0.3899	0.7798
006	9-Vinylanthracene	16	18	21.7898	0.3217	1.0000	0.3435	0.6870
007	1-Vinylphenanthrene	16	18	21.8928	0.3274	0.7203	0.5071	1.0142
008	2-Vinylphenanthrene	16	18	21.8752	0.3264	0.6180	0.5915	1.1830
009	3-Vinylphenanthrene	16	18	21.8782	0.3266	0.7681	0.5234	1.0467
010	4-Vinylphenanthrene	16	18	21.8883	0.3271	0.6753	0.5323	1.0647
011	9-Vinylphenanthrene	16	18	21.8943	0.3275	0.7622	0.4920	0.9840

[a] The energy levels are symmetrically disposed about the $k = 0$ value.

34. Quinodimethanes[a]

No.	Compound	n	m	W	DE_{sp}	k_2	k_1	$E(N \rightarrow V_1)$
001	1,2-Benzoquinodimethane	8	8	9.9540	0.2442	1.1935	0.2950	0.5899
002	1,4-Benzoquinodimethane	8	8	9.9248	0.2406	1.0000	0.3111	0.6222
003	1,4-Naphthoquinodimethane	12	13	15.7999	0.2923	0.8631	0.3595	0.7189
004	1,8-Naphthoquinodimethane	12	13	15.2644	0.2511	1.0000	0.0000	0.0000
005	2,6-Naphthoquinodimethane	12	13	15.4749	0.2673	0.6998	0.1879	0.3757
006	1,4-Phenanthroquinodimethane	16	18	21.5104	0.3061	0.6799	0.3259	0.6519
007	9,10-Phenanthroquinodimethane	16	18	21.6792	0.3155	0.8051	0.4247	0.8495
008	1,4-Anthraquinodimethane	16	18	21.4813	0.3045	0.5559	0.3739	0.7478
009	1,5-Anthraquinodimethane	16	18	21.1709	0.2873	0.6712	0.1027	0.2054
010	2,6-Anthraquinodimethane	16	18	21.0561	0.2809	0.5174	0.1256	0.2511
011	9,10-Anthraquinodimethane	16	18	21.6790	0.3155	0.7574	0.4249	0.8499
012	1,6-Pyrenequinodimethane	18	21	24.5364	0.3113	0.6994	0.2152	0.4303
013	1,8-Pyrenequinodimethane	18	21	24.5387	0.3114	0.7345	0.2131	0.4262
014	4,5-Pyrenequinodimethane	18	21	24.7485	0.3214	0.5581	0.4324	0.8648
015	6,12-Chrysenequinodimethane	20	23	27.2020	0.3131	0.7738	0.2361	0.4723
016	5,12-Naphthacenequinodimethane	20	23	27.3627	0.3201	0.5392	0.4450	0.8901
017	5,11-Naphthacenequinodimethane	20	23	27.2071	0.3134	0.6857	0.2368	0.4737

[a] The energy levels are symmetrically disposed about the k = 0 value.

35. Phenylpolyenes[a]

No.	Compound	n	m	W	DE_{sp}	k_2	k_1	$E(N \rightarrow V_1)$
	Even Systems							
001	Styrene	8	8	10.4243	0.3030	1.0000	0.6622	1.3243
002	1-Phenylbutadiene	10	10	12.9321	0.2932	1.0000	0.4736	0.9472
003	1-Phenylhexatriene	12	12	15.4590	0.2883	1.0000	0.3658	0.7316
004	1-Phenyloctatetraene	14	14	17.9935	0.2853	0.8454	0.2974	0.5947
005	1-Phenyldecapentaene	16	16	20.5319	0.2832	0.7254	0.2503	0.5006
006	1-Phenyldodecahexaene	18	18	23.0726	0.2818	0.6328	0.2160	0.4321
007	1-Phenyltetradecaheptaene	20	20	25.6147	0.2807	0.5600	0.1900	0.3800
008	1-Phenyleicosadecaene	26	26	33.2456	0.2787	0.4149	0.1395	0.2790

[a] The energy levels are symmetrically disposed about the $k = 0$ level.

35. Phenylpolyenes[a]

No.	Compound		n	m	W	DE_{sp}	k_2	k_1	k_{-1}	$E(N \longrightarrow V_1)$
	Odd Systems									
009	Benzyl	C	6	7	8.7206	0.3887	1.2593	1.0000	0.0000	1.0000
010	Cinnamyl	C	8	9	11.3846	0.3761	1.0000	1.0000	0.0000	1.0000
011	1-Phenylpentadienyl	C	10	11	13.9834	0.3621	1.0000	0.7892	0.0000	0.7892
012	1-Phenylheptatrienyl	C	12	13	16.5597	0.3507	1.0000	0.6401	0.0000	0.6401
013	1-Phenylnonatetraenyl	C	14	15	19.1255	0.3417	1.0000	0.5354	0.0000	0.5354
014	1-Phenylundecapentaenyl	C	16	17	21.6855	0.3344	0.8776	0.4590	0.0000	0.4590
015	1-Phenyltridecahexaenyl	C	18	19	24.2420	0.3285	0.7776	0.4013	0.0000	0.4013
016	1-Phenylpentadecaheptaenyl	C	20	21	26.7963	0.3236	0.6961	0.3563	0.0000	0.3563
017	1-Phenylheneicosadecaenyl	C	26	27	34.4510	0.3130	0.5264	0.2663	0.0000	0.2663

[a] The energy levels are symmetrically disposed about the k_{-1} = 0 level. In the case of anions the k_{-1} level of cation is the k_1 level of anion. Values m, W, DE_{sp} and $E(N \longrightarrow V_1)$ are unchanged.

36. Phenylaryls, Biaryls and Polyphenyls[a]

No.	Compound	n	m	W	DE_{sp}	k_2	k_1	$E(N \rightarrow V_1)$
	Phenylaryls, Biaryls							
001	Biphenyl	12	13	16.3834	0.3372	1.0000	0.7046	1.4092
002	1-Phenylnaphthalene	16	18	22.0844	0.3380	1.0000	0.5225	1.0450
003	2-Phenylnaphthalene	16	18	22.0696	0.3372	0.8113	0.5652	1.1305
004	1,1´-Binaphthyl	20	23	27.7891	0.3387	0.7719	0.4433	0.8865
005	1,2´-Binaphthyl	20	23	27.7715	0.3379	0.6896	0.4823	0.9645
006	2,2´-Binaphthyl	20	23	27.7559	0.3372	0.6641	0.5206	1.0412
007	1-Phenylanthracene	20	23	27.7184	0.3356	0.8056	0.3813	0.7625
008	2-Phenylanthracene	20	23	27.7024	0.3349	0.7769	0.3962	0.7924
009	9-Phenylanthracene	20	23	27.7392	0.3365	1.0000	0.3580	0.7160
010	1-Phenylphenanthrene	20	23	27.8482	0.3412	0.7258	0.5297	1.0594
011	2-Phenylphenanthrene	20	23	27.8337	0.3406	0.6404	0.6012	1.2024
012	3-Phenylphenanthrene	20	23	27.8361	0.3407	0.7683	0.5428	1.0856
013	4-Phenylphenanthrene	20	23	27.8445	0.3411	0.6852	0.5535	1.1070
014	9-Phenylphenanthrene	20	23	27.8493	0.3413	0.7631	0.5151	1.0303
	Polyphenyls							
001	Biphenyl	12	13	16.3834	0.3372	1.0000	0.7046	1.4092
015	o-Terphenyl	18	20	24.7776	0.3389	0.8243	0.6103	1.2206
016	m-Terphenyl	18	20	24.7658	0.3383	0.7654	0.6622	1.3243
017	p-Terphenyl	18	20	24.7723	0.3386	1.0000	0.5926	1.1853
018	1,3,5-Triphenylbenzene	24	27	33.1473	0.3388	0.6622	0.6622	1.3243
019	Hexaphenylbenzene	42	48	58.3760	0.3404	0.5047	0.5047	1.0094

36. Phenylaryls, Biaryls and Polyphenyls[a] (cont. 1)

No.	Compound	n	m	W	DE_{sp}	k_2	k_1	$E(N \rightarrow V_1)$
	para-Derivatives							
001	Biphenyl	12	13	16.3834	0.3372	1.0000	0.7046	1.4092
017	p-Terphenyl	18	20	24.7723	0.3386	1.0000	0.5926	1.1853
020	p-Quaterphenyl	24	27	33.1618	0.3393	0.8284	0.5361	1.0723
021	p-Quinquephenyl	30	34	41.5513	0.3397	0.7240	0.5031	1.0062
022	p-Sexiphenyl	36	41	49.9409	0.3400	0.6552	0.4820	0.9640
023	p-Septiphenyl	42	48	58.3304	0.3402	0.6073	0.4677	0.9354
	meta-Derivatives							
001	Biphenyl	12	13	16.3834	0.3372	1.0000	0.7046	1.4092
016	m-Terphenyl	18	20	24.7658	0.3383	0.7654	0.6622	1.3243
024	m-Quaterphenyl	24	27	33.1483	0.3388	0.7151	0.6446	1.2892
025	m-Quinquephenyl	30	34	41.5308	0.3391	0.6851	0.6358	1.2715
026	m-Sexiphenyl	36	41	49.9132	0.3393	0.6668	0.6307	1.2614
027	m-Septiphenyl	42	48	58.2957	0.3395	0.6549	0.6275	1.2550
028	m-Octaphenyl	48	55	66.6782	0.3396	0.6469	0.6254	1.2508

[a] The energy levels are symmetrically disposed about the k = 0 value.

37. Arylphenylmethyl Ions[a]

No.	Compound	n	m	W	DE_{sp}	k_2	k_1	k_{-1}	k_{-2}	$E(N \rightarrow V_1)$
001	Diphenylmethyl	12	14	17.3006	0.3786	1.0000	1.0000	0.0000	-1.0000	1.0000
002	Triphenylmethyl	18	21	25.8003	0.3714	1.0000	1.0000	0.0000	-1.0000	1.0000
003	1-Naphthylphenylmethyl	16	19	23.0518	0.3711	1.0000	0.7141	0.0000	-0.7141	0.7141
004	2-Naphthylphenylmethyl	16	19	23.0004	0.3684	1.0000	0.6534	0.0000	-0.6534	0.6534
005	1-Anthrylphenylmethyl	20	24	28.7095	0.3629	1.0000	0.4960	0.0000	-0.4960	0.4960
006	2-Anthrylphenylmethyl	20	24	28.6499	0.3604	1.0000	0.4569	0.0000	-0.4569	0.4569
007	9-Anthrylphenylmethyl	20	24	28.7906	0.3663	1.0000	0.5778	0.0000	-0.5778	0.5778
008	1-Phenanthrylphenylmethyl	20	24	28.8102	0.3671	0.8052	0.6567	0.0000	-0.6567	0.6567
009	2-Phenanthrylphenylmethyl	20	24	28.7602	0.3650	0.8212	0.6060	0.0000	-0.6060	0.6060
010	3-Phenanthrylphenylmethyl	20	24	28.7737	0.3656	0.7694	0.6600	0.0000	-0.6600	0.6600
011	4-Phenanthrylphenylmethyl	20	24	28.7950	0.3665	0.8211	0.6275	0.0000	-0.6275	0.6275
012	9-Phenanthrylphenylmethyl	20	24	28.8177	0.3674	0.7768	0.6957	0.0000	-0.6957	0.6957
013	2-Pyrenylphenylmethyl	22	27	31.8034	0.3631	0.9147	0.4450	0.0000	-0.4450	0.4450
014	1-Chrysenylphenylmethyl	24	29	34.5597	0.3641	0.7945	0.5733	0.0000	-0.5733	0.5733
015	6-Chrysenylphenylmethyl	24	29	34.5784	0.3648	0.7928	0.6191	0.0000	-0.6191	0.6191
016	4-Diphenylylphenylmethyl	18	21	25.7096	0.3671	1.0000	0.7721	0.0000	-0.7721	0.7721

[a] The energy levels are symmetrically disposed about the k_{-1} = 0 level. In the case of anions the k_{-1} level of cation is the k_1 level of anion. Values m, W, DE_{sp} and $E(N \rightarrow V_1)$ are unchanged.

38. α,ω-Diphenylpolyenes

Even Polyenes[a]

No.	Compound	n	m	W	DE_{sp}	k_2	k_1	$E(N \rightarrow V_1)$
001	Biphenyl	12	13	16.3834	0.3372	1.0000	0.7046	1.4092
002	Stilbene	14	15	18.8778	0.3252	1.0000	0.5043	1.0086
003	1,4-Diphenylbutadiene	16	17	21.4010	0.3177	1.0000	0.3859	0.7718
004	1,6-Diphenylhexatriene	18	19	23.9340	0.3123	0.8584	0.3111	0.6222
005	1,8-Diphenyloctatetraene	20	21	26.4716	0.3082	0.7411	0.2602	0.5204
006	1,10-Diphenyldecapentaene	22	23	29.0118	0.3049	0.6475	0.2235	0.4469
007	1,12-Diphenyldodecahexaene	24	25	31.5535	0.3021	0.5730	0.1958	0.3915
008	1,14-Diphenyltetradecaheptaene	26	27	34.0963	0.2999	0.5129	0.1741	0.3483
009	1,20-Diphenyleicosadecaene	32	33	41.7285	0.2948	0.3887	0.1308	0.2615

[a] The energy levels are symmetrically disposed about the k = 0 value.

38. α,ω-Diphenyl Polyenes (cont. 1)

Odd Polyenes[a]

No.	Compound	n	m	W	DE_{sp}	k_2	k_1	k_{-1}	$E(N \rightarrow V_1)$
010	Diphenyl Methyl	C 12	14	17.3006	0.3786	1.0000	1.0000	0.0000	1.0000
011	1,3-Diphenyl Allyl	C 14	16	19.9105	0.3694	1.0000	0.8106	0.0000	0.8106
012	1,5-Diphenyl Pentadienyl	C 16	18	22.4911	0.3606	1.0000	0.6622	0.0000	0.6622
013	1,7-Diphenyl Heptatrienyl	C 18	20	25.0590	0.3530	1.0000	0.5536	0.0000	0.5536
014	1,9-Diphenyl Nonatetraenyl	C 20	22	27.6203	0.3464	0.8864	0.4736	0.0000	0.4736

[a] The energy levels are symmetrically disposed about the k_{-1} = 0 level. In the case of anions the k_{-1} level of cation is the k_1 level of anion. Values m, W, DE_{sp} and $E(N \rightarrow V_1)$ are unchanged.

41. Benzocyclopentadienyl Ions

No.	Compound		n	m	W	DE_{sp}	k_2	k_1	k_{-1}	k_{-2}	$E(N \rightarrow V_1)$
001	Cyclopentadienyl	C	4		5.2361	0.2472	2.0000	0.6180	0.6180	-1.6180	0.0000
		A	6	5	6.4721	0.4944	0.6180	0.6180	-1.6180	-1.6180	2.2361
002	Indenyl	C	8		11.5808	0.3581	1.1935	0.7293	0.2950	-0.9016	0.4343
		A	10	10	12.1707	0.4171	0.7293	0.2950	-0.9016	-1.2950	1.1966
003	Benz(e)indenyl	C	12		17.2342	0.3489	1.0000	0.5482	0.3557	-0.7435	0.1926
		A	14	15	17.9456	0.3964	0.5482	0.3557	-0.7435	-1.0000	1.0991
004	Benz(f)indenyl	C	12		17.4639	0.3643	0.8435	0.7709	0.1694	-0.5753	0.6016
		A	14	15	17.8027	0.3868	0.7709	0.1694	-0.5753	-1.1694	0.7447
005	Fluorenyl	C	12		17.5437	0.3696	1.0000	0.7046	0.1811	-0.8118	0.5236
		A	14	15	17.9059	0.3937	0.7046	0.1811	-0.8118	-1.0000	0.9929
006	Benzo(a)fluorenyl	C	16		23.2781	0.3639	0.8400	0.6206	0.2005	-0.7399	0.4202
		A	18	20	23.6791	0.3840	0.6206	0.2005	-0.7399	-0.8419	0.9404
007	Benzo(b)fluorenyl	C	16		23.3269	0.3663	0.8449	0.6319	0.1130	-0.5677	0.5189
		A	18	20	23.5528	0.3776	0.6319	0.1130	-0.5677	-0.9239	0.6806
008	Benzo(c)fluorenyl	C	16		23.2180	0.3609	1.0000	0.5291	0.2230	-0.6633	0.3061
		A	18	20	23.6640	0.3832	0.5291	0.2230	-0.6633	-1.0000	0.8863
009	Cyclopenta(c)phenanthryl	A	18		23.6840	0.3842	0.5404	0.3182	-0.6662	-0.7904	0.9844
		C	16	20	23.0476	0.3524	0.8408	0.5404	0.3182	-0.6662	0.2222
010	Cyclopent(a)anthryl	C	16		22.9142	0.3457	0.8249	0.4331	0.3449	-0.5155	0.0882
		A	18	20	23.6040	0.3802	0.4331	0.3449	-0.5155	-0.9446	0.8604

41. Benzocyclopentadienyl Ions (cont. 1)

No.	Compound	n	π	W	DE_{sp}	k_2	k_1	k_{-1}	k_{-2}	$E(N \rightarrow V_1)$
011	Cyclopenta(a)phenanthryl	C 16		23.0711	0.3536	0.7553	0.5928	0.3084	-0.6143	0.2844
		A 18	20	23.6880	0.3844	0.5928	0.3084	-0.6143	-0.9093	0.9227
012	Cyclopenta(1)phenanthryl	C 16		22.9235	0.3462	0.8051	0.4952	0.4247	-0.7596	0.0705
		A 18	20	23.7730	0.3886	0.4952	0.4247	-0.7596	-0.8367	1.1843
013	Cyclopenta(b)phenanthryl	C 16		23.1812	0.3591	0.8456	0.6180	0.2063	-0.5954	0.4117
		A 18	20	23.5938	0.3797	0.6180	0.2063	-0.5954	-0.7986	0.8017
014	Cyclopent(b)anthryl	C 16		23.2026	0.3601	0.7896	0.6180	0.1088	-0.3947	0.5093
		A 18	20	23.4202	0.3710	0.6180	0.1088	-0.3947	-0.9486	0.5035
015	Dibenzo(a,c)fluorenyl	C 20		29.0272	0.3611	0.7827	0.5597	0.2328	-0.7210	0.3268
		A 22	25	29.4929	0.3797	0.5597	0.2328	-0.7210	-0.7768	0.9539
016	Dibenzo(a,g)fluorenyl	C 20		28.9383	0.3575	0.7720	0.4918	0.2504	-0.6699	0.2414
		A 22	25	29.4392	0.3776	0.4918	0.2504	-0.6699	-0.7608	0.9203
017	Dibenzo(a,h)fluorenyl	C 20		29.0764	0.3631	0.6714	0.6484	0.1234	-0.5574	0.5250
		A 22	25	29.3232	0.3729	0.6484	0.1234	-0.5574	-0.7683	0.6808
018	Dibenzo(a,i)fluorenyl	C 20		28.9846	0.3594	0.8592	0.5206	0.2326	-0.6641	0.2880
		A 22	25	29.4497	0.3780	0.5206	0.2326	-0.6641	-0.8555	0.8967
019	Dibenzo(b,g)fluorenyl	C 20		29.0336	0.3613	0.7575	0.5384	0.1356	-0.5226	0.4028
		A 22	25	29.3049	0.3722	0.5384	0.1356	-0.5226	-0.7911	0.6583
020	Dibenzo(b,h)fluorenyl	C 20		29.0614	0.3625	0.8592	0.5206	0.0741	-0.5211	0.4465
		A 22	25	29.2097	0.3684	0.5206	0.0741	-0.5211	-0.6641	0.5953

41. Benzocyclopentadienyl Ions (cont. 2)

No.	Compound		n	m	W	DE_{sp}	k_2	k_1	k_{-1}	k_{-2}	$E(N \to V_1)$
021	Dibenzo(c,g)fluorenyl	C	20		28.8762	0.3550	0.7784	0.4433	0.2766	-0.6441	0.1667
		A	22	25	29.4294	0.3772	0.4433	0.2766	-0.6441	-0.7719	0.9207

41. Benzocyclopentadienyl Ions (cont. 3)

No.	Compound		n	m	W	DE_{sp}	k_2	k_1	k_{-1}	k_{-2}	$E(N \to V_1)$
001	Cyclopentadienyl	C	4		5.2361	0.2472	2.0000	0.6180	0.6180	-1.6180	0.0000
		A	6	5	6.4721	0.4944	0.6180	0.6180	-1.6180	-1.6180	2.2361
002	Indenyl	C	8		11.5808	0.3581	1.1935	0.7293	0.2950	-0.9016	0.4343
		A	10	10	12.1707	0.4171	0.7293	0.2950	-0.9016	-1.2950	1.1966
004	Benz(f)indenyl	C	12		17.4639	0.3643	0.8435	0.7709	0.1694	-0.5753	0.6016
		A	14	15	17.8027	0.3868	0.7709	0.1694	-0.5753	-1.1694	0.7447
014	Cyclopent(b)anthryl	C	16		23.2026	0.3601	0.7896	0.6180	0.1088	-0.3947	0.5093
		A	18	20	23.4202	0.3710	0.6180	0.1088	-0.3947	-0.9486	0.5035
022	Cyclopenta(b)naphthacenyl	C	20		28.8829	0.3553	0.7983	0.4697	0.0753	-0.2853	0.3943
		A	22	25	29.0336	0.3613	0.4697	0.0753	-0.2853	-0.7434	0.3607
023	Cyclopenta(b)pentacenyl	C	24		34.5354	0.3512	0.7413	0.3677	0.0551	-0.2146	0.3127
		A	26	30	34.6456	0.3549	0.3677	0.0551	-0.2146	-0.5957	0.2697
024	Cyclopenta(b)hexacenyl	C	28		40.1733	0.3478	0.6180	0.2950	0.0419	-0.1665	0.2530
		A	30	35	40.2572	0.3502	0.2950	0.0419	-0.1665	-0.4867	0.2085

42. Benzofulvenes

No.	Compound	n	m	W	DE_{sp}	k_2	k_1	k_{-1}	k_{-2}	$E(N \rightarrow V_1)$
001	Fulvene	6	6	7.4659	0.2443	1.0000	0.6180	-0.2541	-1.6180	0.8721
002	Benzo(a)fulvene	10	11	13.3341	0.3031	0.8108	0.5855	-0.3399	-0.9241	0.9254
003	Dibenzofulvene	14	16	19.2237	0.3265	0.7046	0.6387	-0.4301	-0.8396	1.0688
004	Benzo(a)naphtho(2,1-c)-fulvene	18	21	24.9326	0.3301	0.7113	0.5101	-0.3588	-0.7778	0.8689
005	Benzo(a)naphtho(2,3-c)-fulvene	18	21	24.9102	0.3291	0.7104	0.4876	-0.4602	-0.5709	0.9478
006	Benzo(a)naphtho(1,2-c)-fulvene	18	21	24.9331	0.3301	0.6701	0.5168	-0.3593	-0.7412	0.8761
007	4,5-Vinylidenephenanthrene	16	19	22.2930	0.3312	0.7691	0.4839	-0.4421	-0.6052	0.9261
008	Dinaphtho(1,2-a:2',1'-c)-fulvene	22	26	30.6190	0.3315	0.6630	0.4343	-0.3809	-0.5746	0.8152
009	Dinaphtho(2,3-a:2',3'-c)-fulvene	22	26	30.5969	0.3307	0.5206	0.4877	-0.4810	-0.5233	0.9687
010	Dinaphtho(2,1-a:1',2'-c)-fulvene	22	26	30.6475	0.3326	0.6783	0.4433	-0.3033	-0.7569	0.7466

42. Benzofulvenes (cont. 1)

No.	Compound	n	m	W	DE_{sp}	k_2	k_1	k_{-1}	k_{-2}	$E(N—V_1)$
001	Fulvene	6	6	7.4659	0.2443	1.0000	0.6180	−0.2541	−1.6180	0.8721
002	Benzo(a)fulvene	10	11	13.3341	0.3031	0.8108	0.5855	−0.3399	−0.9241	0.9254
011	Naphtho(2,3−a)fulvene	14	16	19.0201	0.3138	0.7875	0.4408	−0.3681	−0.5896	0.8089
012	Anthra(2,3−a)fulvene	18	21	24.6566	0.3170	0.7163	0.3304	−0.3630	−0.4204	0.6934
013	Naphthaceno(2,3−a)fulvene	22	26	30.2770	0.3183	0.6055	0.2523	−0.2829	−0.3924	0.5352
014	Pentaceno(2,3−a)fulvene	26	31	35.8917	0.3191	0.5082	0.1965	−0.2141	−0.3897	0.4106
015	Benzo(b)fulvene	10	11	13.0430	0.2766	1.0000	0.2950	−0.1334	−1.0000	0.4284
016	Naphtho(2,3−b)fulvene	14	16	18.6444	0.2903	0.8435	0.1694	−0.0848	−0.6678	0.2542
017	Anthra(2,3−b)fulvene	18	21	24.2526	0.2977	0.6180	0.1088	−0.0594	−0.4793	0.1682
018	Naphthaceno(2,3−b)fulvene	22	26	29.8630	0.3024	0.4697	0.0753	−0.0441	−0.3618	0.1195
019	Pentaceno(2,3−b)fulvene	26	31	35.4740	0.3056	0.3677	0.0551	−0.0342	−0.2834	0.0892

43. Odd Cyclopentadienylpolyenes and Their Benzo Derivatives

No.	Compound		n	m	W	DE$_{sp}$	k$_2$	k$_1$	k$_{-1}$	k$_{-2}$	E(N$\rightarrow$V$_1$)
001	Vinylcyclopentadienyl	A	8		8.9174	0.4168	0.6180	0.4206	-0.8958	-1.6180	1.3164
		C	6	7	8.0761	0.2966	1.2685	0.6180	0.4206	-0.8958	0.1974
002	Butadienylcyclopentadienyl	A	10		11.4345	0.3816	0.6180	0.3111	-0.6180	-1.4812	0.9291
		C	8	9	10.8123	0.3125	1.0000	0.6180	0.3111	-0.6180	0.3069
003	Hexatrienylcyclopenta-dienyl	A	12		13.9665	0.3606	0.6180	0.2450	-0.4723	-1.1680	0.7172
		C	10	11	13.4766	0.3161	0.8329	0.6180	0.2450	-0.4723	0.3731
004	Octatetraenylcyclopenta-dienyl	A	14		16.5043	0.3465	0.6180	0.2015	-0.3824	-0.9581	0.5839
		C	12	13	16.1014	0.3155	0.7114	0.6180	0.2015	-0.3824	0.4166
005	1-Vinylindenyl	A	12		14.6496	0.3875	0.6963	0.2240	-0.8295	-1.0000	1.0535
		C	10	12	14.2016	0.3501	1.0000	0.6963	0.2240	-0.8295	0.4724
006	1-Butadienylindenyl	A	14		17.1813	0.3701	0.6535	0.1779	-0.6313	-0.9271	0.8092
		C	12	14	16.8255	0.3447	0.8460	0.6535	0.1779	-0.6313	0.4756
007	1-Hexatrienylindenyl	A	16		19.7215	0.3576	0.5945	0.1466	-0.4946	-0.9088	0.6413
		C	14	16	19.4282	0.3393	0.7724	0.5945	0.1466	-0.4946	0.4479
008	1-Octatetraenyllindenyl	A	18		22.2645	0.3480	0.5332	0.1243	-0.4063	-0.8799	0.5306
		C	16	18	22.0159	0.3342	0.7353	0.5332	0.1243	-0.4063	0.4089
009	9-Vinylfluorenyl	A	16		20.4224	0.3778	0.7046	0.1317	-0.7782	-1.0000	0.9099
		C	14	17	20.1590	0.3623	1.0000	0.7046	0.1317	-0.7782	0.5729
010	9-Butadienylfluorenyl	A	18		22.9706	0.3669	0.7046	0.1031	-0.6571	-0.8515	0.7602
		C	16	19	22.7644	0.3560	0.7691	0.7046	0.1031	-0.6571	0.6015

43. Odd Cyclopentadienylpolyenes and Their Benzo Derivatives (cont. 1)

No.	Compound		n	m	W	DE_{sp}	k_2	k_1	k_{-1}	k_{-2}	$E(N \rightarrow V_1)$
011	9-Hexatrienylfluorenyl	A	20	21	25.5204	0.3581	0.6291	0.0846	-0.5284	-0.8245	0.6130
		C	18		25.3512	0.3501	0.7046	0.6291	0.0846	-0.5284	0.5445
012	9-Octatetraenylfluorenyl	A	22	23	28.0697	0.3509	0.5347	0.0717	-0.4391	-0.8072	0.5108
		C	20		27.9263	0.3446	0.7046	0.5347	0.0717	-0.4391	0.4630
013	11-Vinylbenzo(a)fluorenyl	A	20	22	26.1809	0.3719	0.5983	0.1528	-0.6936	-0.8224	0.8464
		C	18		25.8753	0.3580	0.8244	0.5983	0.1528	-0.6936	0.4454
014	11-Butadienylbenzo(a)-fluorenyl	A	22	24	28.7221	0.3634	0.5729	0.1226	-0.5928	-0.7910	0.7154
		C	20		28.4768	0.3532	0.7672	0.5729	0.1226	-0.5928	0.4503
015	11-Hexatrienylbenzo(a)-fluorenyl	A	24	26	31.2676	0.3564	0.5357	0.1021	-0.4874	-0.7653	0.5895
		C	22		31.0635	0.3486	0.6884	0.5357	0.1021	-0.4874	0.4336
016	11-Vinylbenzo(b)fluorenyl	A	20	22	26.0775	0.3672	0.6075	0.0844	-0.5671	-0.8810	0.6516
		C	18		25.9087	0.3595	0.8269	0.6075	0.0844	-0.5671	0.5231
017	11-Butadienylbenzo(b)-fluorenyl	A	22	24	28.6307	0.3596	0.5754	0.0672	-0.5660	-0.6884	0.6332
		C	20		28.4961	0.3540	0.7636	0.5754	0.0672	-0.5660	0.5082
018	11-Hexatrienylbenzo(b)-fluorenyl	A	24	26	31.1839	0.3532	0.5267	0.0557	-0.5392	-0.5735	0.5949
		C	22		31.0724	0.3489	0.6904	0.5267	0.0557	-0.5392	0.4710
019	7-Vinylbenzo(c)fluorenyl	A	20	22	26.1711	0.3714	0.5262	0.1642	-0.6247	-1.0000	0.7889
		C	18		25.8427	0.3565	1.0000	0.5262	0.1642	-0.6247	0.3620
020	7-Butadienylbenzo(c)fluorenyl	A	22	24	28.7139	0.3631	0.5238	0.1293	-0.5613	-0.7934	0.6906
		C	20		28.4553	0.3523	0.7822	0.5238	0.1293	-0.5613	0.3945

43. Odd Cyclopentadienylpolyenes and Their Benzo Derivatives (cont. 2)

No.	Compound		n	m	W	DE_{sp}	k_2	k_1	k_{-1}	k_{-2}	$E(N \rightarrow V_1)$
021	7-Hexatrienylbenzo(c)-fluorenyl	A	24	26	31.2602	0.3562	0.5204	0.1064	-0.4788	-0.7168	0.5853
		C	22		31.0474	0.3480	0.6476	0.5204	0.1064	-0.4788	0.4139
022	4-Vinylcyclopenta(def)-phenanthryl	A	18	20	23.4805	0.3740	0.7056	0.0978	-0.6052	-0.8680	0.7030
		C	16		23.2850	0.3642	0.7691	0.7056	0.0978	-0.6052	0.6079
023	4-Butadienylcyclopenta-(def)phenanthryl	A	20	22	26.0303	0.3650	0.6100	0.0783	-0.6052	-0.6727	0.6836
		C	18		25.8736	0.3579	0.7691	0.6100	0.0783	-0.6052	0.5317
024	4-Hexatrienylcyclopenta-(def)phenanthryl	A	22	24	28.5814	0.3576	0.5307	0.0652	-0.5349	-0.6052	0.6002
		C	20		28.4509	0.3521	0.7691	0.5307	0.0652	-0.5349	0.4655

44. Odd α,ω-Dicyclopentadienylpolyenes and Their Benzo Derivatives

No.	Compound		n	m	W	DE_{sp}	k_2	k_1	k_{-1}	k_{-2}	$E(N \rightarrow V_1)$
001	(Fulven-6-yl)cyclopenta-dienyl	A	12		14.6005	0.3834	0.6180	0.6180	-0.3565	-1.6180	0.9745
		C	10	12	13.3644	0.2804	0.6180	0.6180	0.6180	-0.3565	0.0000
002	Di(fulven-6-yl)methyl	A	14		17.1409	0.3672	0.6180	0.5043	-0.2541	-1.1554	0.7584
		C	12	14	16.1323	0.2952	0.6180	0.6180	0.5043	-0.2541	0.1137
003	1,3-Di(fulven-6-yl)allyl	A	16		19.6821	0.3551	0.6180	0.4206	-0.1982	-0.8958	0.6188
		C	14	16	18.8408	0.3026	0.6180	0.6180	0.4206	-0.1982	0.1974
004	1-(Benzo(a)fulven-8-yl)-indenyl	A	20		26.1927	0.3724	0.6278	0.2950	-0.4486	-0.9016	0.7435
		C	18	22	25.6027	0.3456	0.7293	0.6278	0.2950	-0.4486	0.3328
005	Di(benzo(a)fulven-8-yl)-methyl	A	22		28.7287	0.3637	0.5855	0.2555	-0.3399	-0.8830	0.5954
		C	22	24	28.2177	0.3424	0.7130	0.5855	0.2555	-0.3399	0.3300
006	1,3-Di(benzo(a)fulven-8--yl)allyl	A	24		31.2673	0.3564	0.5410	0.2240	-0.2752	-0.8295	0.4991
		C	22	26	30.8193	0.3392	0.6963	0.5410	0.2240	-0.2752	0.3171
007	9-(Dibenzofulven-10-yl)-fluorenyl	A	28		37.8366	0.3699	0.7046	0.1811	-0.5459	-0.8118	0.7269
		C	26	32	37.4744	0.3586	0.7046	0.7046	0.1811	-0.5459	0.5236
008	Bis(dibenzofulven-10-yl)-methyl	A	30		40.3770	0.3640	0.6387	0.1526	-0.4301	-0.8000	0.5827
		C	28	34	40.0718	0.3551	0.7046	0.6387	0.1526	-0.4301	0.4861
009	1,3-Bis(dibenzofulven-10--yl)allyl	A	32		42.9192	0.3589	0.5490	0.1317	-0.3552	-0.7782	0.4869
		C	30	36	42.6558	0.3516	0.7046	0.5490	0.1317	-0.3552	0.4173

51. Benzotropyls

No.	Compound		n	m	W	DE_{sp}	k_2	k_1	k_{-1}	k_{-2}	$E(N \rightarrow V_1)$
001	Tropyl	C	6		8.9879	0.4268	1.2470	1.2470	-0.4450	-0.4450	1.6920
		A	8	7	8.0978	0.2997	1.2470	-0.4450	-0.4450	-1.8019	0.0000
002	Benzotropyl	C	10		14.7040	0.3920	1.1557	0.8019	-0.2261	-0.5550	1.0281
		A	12	12	14.2518	0.3543	0.8019	-0.2261	-0.5550	-1.0818	0.3288
003	Dibenzo(a,c)tropyl	C	14		20.4664	0.3804	0.8503	0.7962	-0.1323	-0.6180	0.9285
		A	16	17	20.2018	0.3648	0.7962	-0.1323	-0.6180	-0.8932	0.4857
004	Dibenzo(a,d)tropyl	C	14		20.4222	0.3778	1.0000	0.6639	-0.1598	-0.5043	0.8237
		A	16	17	20.1026	0.3590	0.6639	-0.1598	-0.5043	-1.0000	0.3445
005	Naphtho(1,2)tropyl	C	14		20.4740	0.3808	0.9145	0.7031	-0.2773	-0.4450	0.9803
		A	16	17	19.9194	0.3482	0.7031	-0.2773	-0.4450	-0.8879	0.1677
006	Naphtho(2,3)tropyl	C	14		20.3454	0.3733	1.1040	0.5356	-0.1362	-0.5938	0.6717
		A	16	17	20.0731	0.3572	0.5356	-0.1362	-0.5938	-0.7543	0.4576
007	Tribenzotropyl	C	18		26.2326	0.3742	0.8243	0.7191	-0.0858	-0.6103	0.8049
		A	20	22	26.0610	0.3664	0.7191	-0.0858	-0.6103	-0.8510	0.5245
008	Benzo(a)naphtho(2,3-c)-tropyl	C	18		26.1222	0.3692	0.8408	0.5498	-0.0846	-0.6068	0.6344
		A	20	22	25.9530	0.3615	0.5498	-0.0846	-0.6068	-0.6980	0.5222
009	Benzo(a)naphtho(1,2-c)-tropyl	C	18		26.2166	0.3735	0.8716	0.6345	-0.1654	-0.4801	0.7999
		A	20	22	25.8859	0.3584	0.6345	-0.1654	-0.4801	-0.8648	0.3148

51. Benzotropyls (cont. 1)

No.	Compound	n	m	W	DE_sp	k_2	k_1	k_{-1}	k_{-2}	$E(N \rightarrow V_1)$
010	Benzo(a)naphtho(2,1-c)-tropyl	C 18	22	26.2292	0.3741	0.8309	0.6740	-0.1536	-0.5227	0.8276
		A 20		25.9221	0.3601	0.6740	-0.1536	-0.5227	-0.8321	0.3691
011	Benzo(a)naphtho(2,1-d)-tropyl	C 18	22	26.1869	0.3721	0.8639	0.6089	-0.1911	-0.4268	0.8000
		A 20		25.8046	0.3548	0.6089	-0.1911	-0.4268	-0.8589	0.2357
012	Benzo(a)naphtho(2,3-d)-tropyl	C 18	22	26.0707	0.3668	0.8560	0.4979	-0.1030	-0.4878	0.6008
		A 20		25.8647	0.3575	0.4979	-0.1030	-0.4878	-0.7629	0.3848
013	Benzo(a)naphtho(1,2-d)-tropyl	C 18	22	26.1973	0.3726	0.7339	0.6968	-0.1734	-0.4864	0.8702
		A 20		25.8505	0.3568	0.6968	-0.1734	-0.4864	-0.7768	0.3130
014	Cyclohepta(c)phenanthryl	C 18	22	26.2144	0.3734	0.7456	0.6455	-0.2498	-0.4558	0.8953
		A 20		25.7149	0.3507	0.6455	-0.2498	-0.4558	-0.7590	0.2060
015	Cyclohept(a)anthryl	C 18	22	26.1324	0.3697	0.8702	0.4986	-0.2863	-0.3589	0.7849
		A 20		25.5598	0.3436	0.4986	-0.2863	-0.3589	-0.6965	0.0726
016	Cyclohepta(a)phenanthryl	C 18	22	26.2181	0.3735	0.8713	0.5927	-0.2456	-0.4759	0.8383
		A 20		25.7268	0.3512	0.5927	-0.2456	-0.4759	-0.7045	0.2303
017	Cyclohepta(1)phenanthryl	C 18	22	26.2987	0.3772	0.7823	0.7459	-0.3264	-0.4033	1.0724
		A 20		25.6458	0.3475	0.7459	-0.3264	-0.4033	-0.7875	0.0769
018	Cyclohepta(b)phenanthryl	C 18	22	26.1331	0.3697	0.7683	0.5622	-0.1649	-0.5280	0.7271
		A 20		25.8033	0.3547	0.5622	-0.1649	-0.5280	-0.7395	0.3631
019	Cyclohept(b)anthryl	C 18	22	25.9666	0.3621	0.8874	0.3766	-0.0905	-0.5542	0.4672
		A 20		25.7856	0.3539	0.3766	-0.0905	-0.5542	-0.6087	0.4637

51. Benzotropyls (cont. 2)

No.	Compound		n	m	W	DE_{sp}	k_2	k_1	k_{-1}	k_{-2}	$E(N \rightarrow V_1)$
001	Tropyl	C	6		8.9879	0.4268	1.2470	1.2470	-0.4450	-0.4450	1.6920
		A	8	7	8.0978	0.2997	1.2470	-0.4450	-0.4450	-1.8019	0.0000
002	Benzotropyl	C	10		14.7040	0.3920	1.1557	0.8019	-0.2261	-0.5550	1.0281
		A	12	12	14.2518	0.3543	0.8019	-0.2261	-0.5550	-1.0818	0.3288
006	Naphtho(2,3)tropyl	C	14		20.3454	0.3733	1.1040	0.5356	-0.1362	-0.5938	0.6717
		A	16	17	20.0731	0.3572	0.5356	-0.1362	-0.5938	-0.7543	0.4576
019	Cyclohept(b)anthryl	C	18		25.9666	0.3621	0.8874	0.3766	-0.0905	-0.5542	0.4672
		A	20	22	25.7856	0.3539	0.3766	-0.0905	-0.5542	-0.6087	0.4637
020	Cyclohepta(b)naphthacenyl	C	22		31.5814	0.3549	0.7069	0.2763	-0.0644	-0.4237	0.3407
		A	24	27	31.4527	0.3501	0.2763	-0.0644	-0.4237	-0.6144	0.3593
021	Cyclohepta(b)pentacenyl	C	26		37.1940	0.3498	0.5728	0.2098	-0.0480	-0.3339	0.2578
		A	28	32	37.0980	0.3468	0.2098	-0.0480	-0.3339	-0.6167	0.2859
022	Cyclohepta(b)hexacenyl	C	30		42.8058	0.3461	0.4718	0.1638	-0.0371	-0.2695	0.2009
		A	32	37	42.7315	0.3441	0.1638	-0.0371	-0.2695	-0.5790	0.2324

52. Benzoheptafulvenes

No.	Compound	n	m	W	DE_{sp}	k_2	k_1	k_{-1}	k_{-2}	$E(N \rightarrow V_1)$
001	Heptafulvene	8	8	9.9944	0.2493	1.2470	0.2163	-0.4450	-0.7764	0.6613
002	Benzo(a)heptafulvene	12	13	15.8589	0.2968	0.8715	0.2813	-0.4375	-0.7262	0.7189
003	Benzo(b)heptafulvene	12	13	15.6021	0.2771	0.8928	0.1172	-0.2297	-0.7957	0.3469
004	Benzo(c)heptafulvene	12	13	15.8361	0.2951	0.8019	0.2936	-0.5509	-0.5550	0.8445
005	Dibenzo(a,c)heptafulvene	16	18	21.7268	0.3182	0.8071	0.3571	-0.4922	-0.6588	0.8493
006	Dibenzo(a,d)heptafulvene	16	18	21.4464	0.3026	0.8054	0.1525	-0.2254	-0.7454	0.3779
007	Dibenzo(a,e)heptafulvene	16	18	21.7284	0.3182	0.7528	0.3581	-0.5043	-0.6180	0.8624
008	Dibenzo(b,d)heptafulvene	16	18	21.2886	0.2938	0.7962	0.0638	-0.1323	-0.7410	0.1961
009	Tribenzoheptafulvene	20	23	27.6179	0.3312	0.7752	0.4331	-0.5709	-0.6103	1.0041

52. Benzoheptafulvenes (cont. 1)

No.	Compound	n	m	W	DE_{sp}	k_2	k_1	k_{-1}	k_{-2}	$E(N \rightarrow V_1)$
001	Heptafulvene	8	8	9.9944	0.2493	1.2470	0.2163	-0.4450	-0.7764	0.6613
002	Benzo(a)heptafulvene	12	13	15.8589	0.2968	0.8715	0.2813	-0.4375	-0.7262	0.7189
010	Naphtho(2,3-a)heptafulvene	16	18	21.5457	0.3081	0.5731	0.3003	-0.3720	-0.6730	0.6723
011	Anthra(2,3-a)heptafulvene	20	23	27.1831	0.3123	0.4091	0.2984	-0.3017	-0.5773	0.6002
012	Naphthaceno(2,3-a)hepta-fulvene	24	28	32.8040	0.3144	0.3393	0.2631	-0.2398	-0.5002	0.5029
013	Pentaceno(2,3-a)heptafulvene	28	33	38.4190	0.3157	0.3247	0.2071	-0.1906	-0.4402	0.3977
003	Benzo(b)heptafulvene	12	13	15.6021	0.2771	0.8928	0.1172	-0.2297	-0.7957	0.3469
014	Naphtho(2,3-b)heptafulvene	16	18	21.2140	0.2897	0.6209	0.0756	-0.1381	-0.7361	0.2136
015	Anthra(2,3-b)heptafulvene	20	23	26.8259	0.2968	0.4535	0.0534	-0.0916	-0.5557	0.1450
016	Naphthaceno(2,3-b)hepta-fulvene	24	28	32.4376	0.3013	0.3453	0.0401	-0.0650	-0.4263	0.1050
017	Pentaceno(2,3-b)heptafulvene	28	33	38.0491	0.3045	0.2717	0.0312	-0.0484	-0.3362	0.0796
004	Benzo(c)heptafulvene	12	13	15.8361	0.2951	0.8019	0.2936	-0.5509	-0.5550	0.8445
018	Naphtho(2,3-c)heptafulvene	16	18	21.5219	0.3068	0.5356	0.3218	-0.4142	-0.5938	0.7360
019	Anthra(2,3-c)heptafulvene	20	23	27.1597	0.3113	0.3766	0.3323	-0.3163	-0.6087	0.6486
020	Naphthaceno(2,3-c)hepta-fulvene	24	28	32.7810	0.3136	0.3363	0.2763	-0.2448	-0.5777	0.5211
021	Pentaceno(2,3-c)hepta-fulvene	28	33	38.3961	0.3150	0.3379	0.2098	-0.1923	-0.4867	0.4021

53001−53007

53. Polyenylcycloheptatrienyl and Vinylbenzotropyl Ions

No.	Compound		n	m	W	DE_{sp}	k_2	k_1	k_{-1}	k_{-2}	$E(N \rightarrow V_1)$
001	Tropyl	C	6	7	8.9879	0.4268	1.2470	1.2470	-0.4450	-0.4450	1.6920
		A	8		8.0978	0.2997	1.2470	-0.4450	-0.4450	-1.8019	0.0000
002	Vinyltropyl	C	8	9	11.4391	0.3821	1.2470	0.8072	-0.3379	-0.4450	1.1451
		A	10		10.7632	0.3070	0.8072	-0.3379	-0.4450	-1.1019	0.1071
003	1-Butadienyltropyl	C	10	11	13.9587	0.3599	1.2470	0.5743	-0.2668	-0.4450	0.8411
		A	12		13.4250	0.3114	0.5743	-0.2668	-0.4450	-0.8494	0.1782
004	1-Hexatrienyltropyl	C	12	13	16.4919	0.3455	1.0975	0.4450	-0.2183	-0.4450	0.6634
		A	14		16.0552	0.3119	0.4450	-0.2183	-0.4450	-0.7113	0.2267
005	1-Vinylbenzo(b)tropyl	C	12	14	17.1834	0.3702	1.0000	0.7044	-0.1825	-0.5211	0.8869
		A	14		16.8184	0.3442	0.7044	-0.1825	-0.5211	-1.0000	0.3386
006	1-Vinylbenzo(c)tropyl	C	12	14	17.1443	0.3675	1.0000	0.6584	-0.2253	-0.4331	0.8837
		A	14		16.6938	0.3353	0.6584	-0.2253	-0.4331	-1.0000	0.2078
007	1-Vinylbenzo(d)tropyl	C	12	14	17.1728	0.3695	0.8288	0.8019	-0.1815	-0.5550	0.9834
		A	14		16.8099	0.3436	0.8019	-0.1815	-0.5550	-0.8888	0.3735

61. Benzopentalenes

No.	Compound	n	m	W	DE_{sp}	k_2	k_1	k_{-1}	k_{-2}	$E(N \rightarrow V_1)$
001	Pentalene	8		10.4556	0.2728	1.0000	0.4707	0.0000	−1.4142	0.4707
		diA 10	9	10.4556	0.4951	0.4707	0.0000	−1.4142	−1.8136	1.4142
002	Benzopentalene	12	14	16.3038	0.3074	0.8120	0.4563	−0.0407	−0.9634	0.4970
003	Dibenzo(a,f)pentalene	16		22.0037	0.3160	0.8108	0.2962	0.0000	−0.9231	0.2962
		diA 18	19	22.0037	0.4212	0.2962	0.0000	−0.9231	−1.0000	0.9231
004	Dibenzo(a,e)pentalene	16	19	22.1870	0.3256	0.5982	0.5441	−0.1157	−0.8474	0.6598

62. Benzazulenes

No.	Compound	n	m	W	DE_{sp}	k_2	k_1	k_{-1}	k_{-2}	$E(N \rightarrow V_1)$
001	Azulene	10	11	13.3635	0.3058	0.8870	0.4773	-0.4004	-0.7376	0.8777
002	Benz(a)azulene	14	16	19.0949	0.3184	0.8417	0.3233	-0.3884	-0.6678	0.7118
003	Benz(e)azulene	14	16	19.1086	0.3193	0.7054	0.4913	-0.2611	-0.7920	0.7524
004	Benz(f)azulene	14	16	19.0837	0.3177	0.8182	0.4219	-0.3165	-0.6535	0.7385
005	Dibenz(a,e)azulene	18	21	24.8825	0.3277	0.6927	0.3680	-0.2977	-0.6847	0.6657
006	Dibenz(a,g)azulene	18	21	24.8591	0.3266	0.6933	0.3500	-0.3651	-0.5650	0.7152
007	Dibenz(a,h)azulene	18	21	24.7964	0.3236	0.7448	0.2829	-0.2212	-0.7268	0.5041
008	Dibenz(a,f)azulene	18	21	24.7781	0.3228	0.8497	0.2595	-0.2548	-0.6655	0.5143

62. Benzazulenes (cont. 1)

No.	Compound	n	m	W	DE_{sp}	k_2	k_1	k_{-1}	k_{-2}	$E(N \rightarrow V_1)$
001	Azulene	10	11	13.3635	0.3058	0.8870	0.4773	-0.4004	-0.7376	0.8777
002	Benz(a)azulene	14	16	19.0949	0.3184	0.8417	0.3233	-0.3884	-0.6678	0.7118
009	Naphth(2,3-a)azulene	18	21	24.7450	0.3212	0.7365	0.2332	-0.3589	-0.5732	0.5921
010	Azulen(1,2-b)anthracene	22	26	30.3706	0.3219	0.5792	0.1784	-0.3059	-0.4843	0.4843
011	Azuleno(1,2-b)naphthacene	26	31	35.9874	0.3222	0.4600	0.1418	-0.2442	-0.4366	0.3860
012	Azuleno(1,2-b)pentacene	30	36	41.6010	0.3222	0.3745	0.1155	-0.1925	-0.4057	0.3080
003	Benz(e)azulene	14	16	19.1086	0.3193	0.7054	0.4913	-0.2611	-0.7920	0.7524
013	Naphth(2,3-e)azulene	18	21	24.7621	0.3220	0.5806	0.4388	-0.1949	-0.7217	0.6337
014	Azulen(4,5-b)anthracene	22	26	30.3884	0.3226	0.5461	0.3361	-0.1547	-0.5376	0.4908
015	Azuleno(4,5-b)naphthacene	26	31	36.0054	0.3228	0.5261	0.2541	-0.1267	-0.4191	0.3808
016	Azuleno(4,5-b)pentacene	30	36	41.6190	0.3228	0.4914	0.1960	-0.1059	-0.3392	0.3019
004	Benz(f)azulene	14	16	19.0837	0.3177	0.8182	0.4219	-0.3165	-0.6535	0.7385
017	Naphth(2,3-f)azulene	18	21	24.7335	0.3206	0.6518	0.3642	-0.2357	-0.6223	0.5999
018	Azulen(5,6-b)anthracene	22	26	30.3592	0.3215	0.5399	0.2984	-0.1819	-0.5447	0.4803
019	Azuleno(5,6-b)naphthacene	26	31	35.9762	0.3218	0.4747	0.2369	-0.1448	-0.4488	0.3816
020	Azuleno(5,6-b)pentacene	30	36	41.5898	0.3219	0.4264	0.1877	-0.1178	-0.3709	0.3055

63. Benzoheptalenes

No.	Compound	n	m	W	DE_{sp}	k_2	k_1	k_{-1}	k_{-2}	$E(N \rightarrow V_1)$
001	Heptalene	12		15.6182	0.2783	1.0000	0.0000	-0.3111	-0.7046	0.3111
	dic	10	13	15.6182	0.4322	1.3174	1.0000	0.0000	-0.3111	1.0000
002	Benzo(a)heptalene	16	18	21.4529	0.3029	0.8788	0.0266	-0.3384	-0.5774	0.3650
003	Benzo(b)heptalene	16	18	21.4090	0.3005	0.7653	0.0313	-0.2840	-0.6633	0.3152
004	Dibenzo(a,c)heptalene	20	23	27.3104	0.3178	0.7800	0.0768	-0.3391	-0.6210	0.4158
005	Dibenzo(a,i)heptalene	20	23	27.2791	0.3165	0.7347	0.0838	-0.3531	-0.5214	0.4369
006	Dibenzo(b,i)heptalene	20	23	27.0971	0.3086	0.7261	0.0000	-0.1840	-0.7105	0.1840
007	Dibenzo(b,h)heptalene	20	23	27.2440	0.3150	0.6641	0.0940	-0.3381	-0.5206	0.4321
008	Dibenzo(a,h)heptalene	20	23	27.1424	0.3105	0.7713	0.0000	-0.2112	-0.6180	0.2112
009	Dibenzo(a,d)heptalene	20	23	27.2042	0.3132	0.8005	0.0000	-0.3371	-0.4678	0.3371
010	Dibenzo(a,j)heptalene	20	23	27.1790	0.3121	0.7843	0.0000	-0.2471	-0.5536	0.2471
011	Dibenzo(a,g)heptalene	20	23	27.3091	0.3178	0.7719	0.0760	-0.4205	-0.4433	0.4965

71. Cata-condensed Tricyclic Hydrocarbons

No.	Compound		n	m	W	DE_{sp}	k_2	k_1	k_{-1}	k_{-2}	$E(N \rightarrow V_1)$
001	Cyclopenta(a)pentalenyl	A	12	13	14.8963	0.3766	0.6180	0.2541	−0.2197	−1.4955	0.4738
002	Cyclopent(a)azulenyl	A	14	15	17.5138	0.3676	0.5714	0.0959	−0.4650	−0.7596	0.5609
		C	12	15	17.3220	0.3548	0.9278	0.5714	0.0959	−0.4650	0.4756
003	Cyclopent(e)azulenyl	A	14	15	17.6782	0.3785	0.5149	0.3320	−0.4041	−1.0000	0.7362
004	Cyclopent(f)azulenyl	A	14	15	17.6743	0.3783	0.6180	0.2752	−0.4939	−0.8350	0.7692
005	Cyclopenta(a)heptalenyl	C	14	17	20.1024	0.3590	0.8083	0.5563	−0.0695	−0.4450	0.6257
		A	16	17	19.9635	0.3508	0.5563	−0.0695	−0.4450	−0.6816	0.3756
006	Cyclopenta(b)heptalenyl	C	14	17	20.0651	0.3568	0.9132	0.4874	−0.0814	−0.3803	0.5688
		A	16	17	19.9024	0.3472	0.4874	−0.0814	−0.3803	−0.7906	0.2989
007	Cyclohept(a)azulenyl	C	14	17	20.2094	0.3653	1.0000	0.4641	−0.2373	−0.4450	0.7014
008	Cyclohepta(a)heptalenyl	C	16	19	22.5657	0.3456	1.0691	0.1331	−0.1749	−0.3912	0.3081
009	Cyclohepta(b)heptalenyl	C	16	19	22.5522	0.3449	0.9106	0.1565	−0.1455	−0.4450	0.3020

72. Indacene-like Compounds

No.	Compound	n	m	W	DE_{sp}	k_2	k_1	k_{-1}	k_{-2}	$E(N \rightarrow V_1)$
001	s-Indacene	12		16.2314	0.3022	0.6180	0.6180	0.0000	-0.8202	0.6180
		diA	14	16.2314	0.4451	0.6180	0.0000	-0.8202	-1.6180	0.8202
002	as-Indacene	12		15.8987	0.2785	0.8323	0.3046	0.2391	-1.0000	0.0655
		diA	14	16.3769	0.4555	0.3046	0.2391	-1.0000	-1.4142	1.2391
003	Cyclohept(f)indene	14	16	18.8942	0.3059	0.8402	0.2607	-0.2154	-0.6607	0.4761
004	Cyclohept(e)indene	14	16	19.0377	0.3149	0.6680	0.4223	-0.3446	-0.5440	0.7668
005	Dicyclohepta(a,d)benzene	16		21.3381	0.2966	0.6796	0.0000	-0.4450	-0.4450	0.4450
		diC	18	21.3381	0.4077	1.2470	0.6796	0.0000	-0.4450	0.6796
006	Dicyclohepta(a,c)benzene	16		21.0970	0.2832	0.8402	-0.1812	-0.2154	-0.6180	0.0342
		diC	18	21.4593	0.4144	1.0685	0.8402	-0.1812	-0.2154	1.0213

72. Indacene-like Compounds (cont. 1)

No.	Compound	n	m	W	DE_{sp}	k_2	k_1	k_{-1}	k_{-2}	$E(N \rightarrow V_1)$
003	Cyclohept(f)indene	14	16	18.8942	0.3059	0.8402	0.2607	-0.2154	-0.6607	0.4761
007	Cyclopenta(b)cyclohepta(g)-naphthalene	18	21	24.4772	0.3084	0.7380	0.1587	-0.1327	-0.6338	0.2914
008	Cyclopenta(b)cyclohept(i)-anthracene	22	26	30.0785	0.3107	0.5596	0.1048	-0.0892	-0.5298	0.1940
009	Cyclopenta(b)cyclohepta(k)-naphthacene	26	31	35.6863	0.3125	0.4359	0.0736	-0.0638	-0.4095	0.1374
004	Cyclohept(c)indene	14	16	19.0377	0.3149	0.6680	0.4223	-0.3446	-0.5440	0.7668
010	Cyclopenta(a)cyclohepta(g)-naphthalene	18	21	24.6630	0.3173	0.5115	0.3892	-0.1996	-0.5859	0.5888
011	Cyclopenta(a)cyclohept(i)-anthracene	22	26	30.2775	0.3184	0.4307	0.3336	-0.1271	-0.6015	0.4607
012	Cyclopenta(a)cyclohepta(k)-naphthacene	26	31	35.8896	0.3190	0.4085	0.2604	-0.0869	-0.4932	0.3473
013	Cyclopenta(b)cyclohepta(f)-naphthalene	18	21	24.6547	0.3169	0.7183	0.2430	-0.3180	-0.4368	0.5609
014	Cyclopenta(b)cyclohepta(h)-phenanthrene	22	26	30.3077	0.3195	0.5558	0.2342	-0.1908	-0.5141	0.4251
015	Cyclohept(b)inden(h)anthra-cene	26	31	35.9318	0.3204	0.4093	0.2299	-0.1233	-0.5486	0.3532
016	Cyclohept(b)indeno(j)-naphthacene	30	36	41.5472	0.3208	0.3052	0.2273	-0.0851	-0.4733	0.3124
017	Cyclopenta(b)cyclohept(h)-anthracene	22	26	30.2637	0.3178	0.6778	0.1518	-0.2958	-0.3567	0.4477

72. Indacene-like Compounds (cont. 2)

No.	Compound	n	m	W	DE_{sp}	k_2	k_1	k_{-1}	k_{-2}	$E(N \rightarrow V_1)$
018	Cyclohept(b)indeno(5,6-h)-phenanthrene	26	31	35.9266	0.3202	0.5698	0.1478	-0.1862	-0.4073	0.3340
019	Cyclopenta(b)cyclohepta(n)-pentaphene	30	36	41.5551	0.3210	0.4230	0.1456	-0.1210	-0.4336	0.2666
020	Cyclopenta(b)cyclohepta(p)-hexaphene	34	41	47.1723	0.3213	0.3139	0.1443	-0.0839	-0.4322	0.2283
021	Cyclopenta(b)cyclohepta(j)-naphthacene	26	31	35.8730	0.3185	0.5311	0.1019	-0.2500	-0.3223	0.3519
022	2,3-Cyclohepta-10,11-cyclo-pentabenzo(a)naphthacene	30	36	41.5394	0.3205	0.5095	0.0999	-0.1838	-0.3074	0.2837
023	Cyclopenta(p)cyclohepta(b)-hexaphene	34	41	47.1698	0.3212	0.4244	0.0986	-0.1197	-0.3223	0.2183

73. Peri-condensed Tricyclic Hydrocarbons

No.	Compound		n	m	W	DE$_{sp}$	k$_2$	k$_1$	k$_{-1}$	k$_{-2}$	E(N→V$_1$)
001	Acepentylene	-	-	10	12.9952	0.2496	0.5720	0.3473	0.3473	-1.0000	0.0000
		diA	12	12	13.6898	0.4742	0.3473	0.3473	-1.0000	-1.8794	1.3473
002	Cyclopent(cd)indenyl	C	10	10	14.3344	0.3334	0.8115	0.6180	0.4619	-0.6180	0.1561
		A	12	13	15.2582	0.4045	0.6180	0.4619	-0.6180	-1.2134	1.0800
003	Cyclopent(cd)azulene	-	12	14	16.3660	0.3119	0.8308	0.4805	-0.2846	-0.5939	0.7651
004	Acenaphthylene	-	12	14	16.6189	0.3299	0.8308	0.6375	-0.2846	-1.0000	0.9221
005	Benz(cd)azulenyl	C	12	12	17.5742	0.3716	1.0000	0.6801	0.0000	-0.5054	0.6801
		A	14	15	17.5742	0.3716	0.6801	0.0000	-0.5054	-1.0000	0.5054
006	Aceheptylene	-	14	16	18.9112	0.3069	0.7580	0.2411	-0.3292	-0.7092	0.5703
007	Phenalenyl	C	12	12	17.8272	0.3885	1.0000	1.0000	0.0000	-1.0000	1.0000
		A	14	15	17.8272	0.3885	1.0000	0.0000	-1.0000	-1.0000	1.0000
008	Pleiadiene	-	14	16	19.1448	0.3215	1.0000	0.2411	-0.4574	-0.7092	0.6985
009	Benzo(ef)heptalenyl	C	14	14	20.3054	0.3709	1.1066	0.4450	-0.3056	-0.4450	0.7506
		A	16	17	19.6942	0.3350	0.4450	-0.3056	-0.4450	-0.6350	0.1395
010	Cyclohepta(ef)heptalene	-	16	16	20.9133	0.2730	0.6180	-0.2091	-0.2091	-0.4142	0.0000
		diC	14	18	21.3315	0.4073	1.3028	0.6180	-0.2091	-0.2091	0.8271

74. Fluoranthene-like Compounds

No.	Compound	n	m	W	DE$_{sp}$	k$_2$	k$_1$	k$_{-1}$	k$_{-2}$	E(N→V$_1$)
001	Fluoranthene	16	19	22.5001	0.3421	0.7436	0.6180	-0.3709	-0.9046	0.9889
002	Benzo(a)fluoranthene	20	24	28.1636	0.3401	0.7181	0.4514	-0.2521	-0.8444	0.7035
003	Benzo(b)fluoranthene	20	24	28.2635	0.3443	0.6874	0.6021	-0.3770	-0.7385	0.9791
004	Benzo(ghi)fluoranthene	18	22	25.6054	0.3457	0.7614	0.6180	-0.3783	-0.6719	0.9964
005	Benzo(j)fluoranthene	20	24	28.2118	0.3422	0.6733	0.5253	-0.3116	-0.7689	0.8369
006	Benzo(k)fluoranthene	20	24	28.1863	0.3411	0.7834	0.4593	-0.4007	-0.5788	0.8600
007	Dibenzo(a,f)fluoranthene	24	29	33.8221	0.3387	0.6096	0.3720	-0.1604	-0.7508	0.5324
008	Dibenzo(a,k)fluoranthene	24	29	33.8488	0.3396	0.6857	0.3678	-0.2752	-0.5736	0.6430
009	Dibenzo(b,ghi)fluoranthene	22	27	31.3558	0.3465	0.7669	0.5129	-0.3922	-0.5849	0.9051
010	Dibenzo(b,k)fluoranthene	24	29	33.9498	0.3431	0.6840	0.4611	-0.4062	-0.5719	0.8673
011	Dibenzo(b,j)fluoranthene	24	29	33.9748	0.3440	0.6319	0.5198	-0.3177	-0.7093	0.8375
012	Dibenzo(j,l)fluoranthene	24	29	33.9976	0.3447	0.6688	0.5049	-0.2846	-0.7602	0.7895
013	Naphtho(2,3-b)fluoranthene	24	29	33.9190	0.3420	0.6707	0.4552	-0.3359	-0.5742	0.7910
014	Naphtho(8,2,1-bcd)fluoranthene	22	27	31.3319	0.3456	0.7116	0.4746	-0.2933	-0.7969	0.7679
015	Naphtho(1,2,3,4-ghi)fluor-anthene	22	27	31.4164	0.3488	0.7153	0.6045	-0.3705	-0.6977	0.9750
016	Naphtho(2,3-j)fluoranthene	24	29	33.8503	0.3397	0.6795	0.3816	-0.2666	-0.5784	0.6482
017	Naphtho(1,2-k)fluoranthene	24	29	33.9532	0.3432	0.6413	0.4857	-0.3749	-0.6039	0.8606
018	Naphtho(2,3-k)fluoranthene	24	29	33.8222	0.3387	0.7378	0.3420	-0.3966	-0.4115	0.7386
019	Tribenzo(a,f,j)fluoranthene	28	34	39.5486	0.3397	0.5017	0.3970	-0.1357	-0.6993	0.5327
020	Fluoreno(2,3,4,9-defg)chrysene	26	32	37.0760	0.3461	0.6935	0.4074	-0.2300	-0.7937	0.6374
021	Benzo(a)naphtho(2,3-j)fluor-anthene	28	34	39.5204	0.3388	0.4941	0.3782	-0.1942	-0.5199	0.5724

74. Fluoranthene-like Compounds (cont. 1)

No.	Compound	n	m	W	DE$_{sp}$	k$_2$	k$_1$	k$_{-1}$	k$_{-2}$	E(N→V$_1$)
022	Benzo(a)naphtho(2,3-k)fluoranthene	28	34	39.4857	0.3378	0.5887	0.2947	-0.2829	-0.3972	0.5776
023	Benzo(a)naphtho(2,3-ℓ)fluoranthene	28	34	39.5241	0.3389	0.5382	0.3561	-0.1917	-0.5328	0.5478
024	Dibenzo(b,e)naphtho(2,1,8-hij)acephenanthrylene	30	37	42.7558	0.3448	0.51407	0.3770	-0.1658	-0.6426	0.5428
025	peri-Phenylenefluoranthene	22	27	31.2842	0.3439	0.6180	0.5798	-0.1859	-0.8549	0.7657
026	Indeno(1,2,3,4-ghij)fluoranthene	20	25	28.6054	0.3442	0.6417	0.6020	-0.1850	-0.7456	0.7869
027	Isorubicene	26	32	36.9646	0.3426	0.5947	0.4453	-0.0880	-0.8007	0.5333
028	Rubicene	26	32	36.9996	0.3437	0.7086	0.4795	-0.1250	-0.7395	0.6045
029	Fluoreno(9,1-ab)fluoranthene	26	32	37.0765	0.3461	0.6708	0.6009	-0.2155	-0.6472	0.8164
030	Acenaphtho(1,2-k)fluoranthene	26	32	37.0128	0.3442	0.6375	0.4911	-0.2846	-0.5366	0.7757
031	Acenaphtho(1,2-j)fluoranthene	26	32	37.0170	0.3443	0.5748	0.5363	-0.3455	-0.3830	0.8817
032	Diindeno(1,2,3-fg:1',2',3'-op)naphthacene	30	37	42.6359	0.3415	0.4707	0.4142	0.0000	-0.7321	0.4142
033	Benzo(b)rubicene	30	37	42.6560	0.3421	0.6104	0.3938	-0.0577	-0.6386	0.4515
034	Diindeno(1,2,3-cd:1',2',3'-jk)pyrene	28	35	40.1424	0.3469	0.6130	0.4982	-0.1766	-0.7247	0.6748
035	Periflanthene	32	40	45.9064	0.3477	0.6225	0.4319	-0.1176	-0.6801	0.5495
036	Benzo(hij)periflanthene	34	43	49.0652	0.3504	0.6345	0.4950	-0.1920	-0.6085	0.6870
037	Naphtho(2,3-j)acenaphtho(1,2-ℓ)fluoranthene	34	42	48.3977	0.3428	0.6116	0.3493	-0.2216	-0.3892	0.5709
038	Dibenzo(g,u)periflanthene	40	50	57.3178	0.3464	0.5135	0.3305	-0.0099	-0.5725	0.3404
039	Decacyclene	36	45	51.5476	0.3455	0.4807	0.4807	-0.3328	-0.3328	0.8136
040	Tetreacenaphth(1,2-a:1',2'-c:1",2"-h:1‴,2‴-j)anthracene	54	68			0.5363	0.3133	-0.1774	-0.3455	0.4907

75. Benzacenaphthylenes

No.	Compound	n	m	W	DE_sp	k_2	k_1	k_{-1}	k_{-2}	$E(N \rightarrow V_1)$
001	Acenaphthylene	12	14	16.6189	0.3299	0.8308	0.6375	-0.2846	-1.0000	0.9221
002	Aceanthrylene	16	19	22.3005	0.3316	0.7119	0.5166	-0.1866	-0.9220	0.7032
003	Benz(k)aceanthrylene	20	24	27.9336	0.3306	0.6797	0.3830	-0.1386	-0.6443	0.5216
004	Naphth(2,3-k)aceanthrylene	24	29	33.5533	0.3294	0.6363	0.2866	-0.1093	-0.4783	0.3959
005	Cyclopenta(de)hexacene	28	34	39.1682	0.3285	0.5543	0.2196	-0.0894	-0.3719	0.3090
006	Cyclopenta(de)heptacene	32	39	44.7813	0.3277	0.4690	0.1723	-0.0749	-0.2995	0.2472
007	Cyclopent(de)octaene	36	44	50.3935	0.3271	0.3969	0.1381	-0.0639	-0.2477	0.2020
008	Acephenanthrylene	16	19	22.3813	0.3359	0.7651	0.5864	-0.2913	-0.7491	0.8777
009	Benz(k)acephenanthrylene	20	24	28.0385	0.3349	0.6811	0.4592	-0.2648	-0.5641	0.7241
010	Naphth(2,3-k)acephenanthrylene	24	29	33.6659	0.3333	0.6450	0.3404	-0.2265	-0.4540	0.5669
011	Indeno(7,1-ab)pentacene	28	34	39.2833	0.3319	0.5958	0.2565	-0.1870	-0.3888	0.4434
012	Indeno(7,1-ab)hexacene	32	39	44.8970	0.3307	0.5176	0.1980	-0.1526	-0.3452	0.3506
013	Indeno(7,1-ab)heptacene	36	44	50.5093	0.3298	0.4386	0.1565	-0.1249	-0.3109	0.2814

76. Fluoranthenopolyacenes

No.	Compound	n	m	W	DE_{sp}	k_2	k_1	k_{-1}	k_{-2}	$E(N \rightarrow V_1)$
001	Fluoranthene	16	19	22.5001	0.3421	0.7436	0.6180	-0.3709	-0.9046	0.9889
002	Benzo(k)fluoranthene	20	24	28.1863	0.3411	0.7834	0.4593	-0.4007	-0.5788	0.8600
003	Naphtho(2,3-k)fluoranthene	24	29	33.8222	0.3387	0.7378	0.3420	-0.3966	-0.4115	0.7386
004	Acenaphthyleno(1,2-b)naphthacene	28	34	39.4421	0.3365	0.6180	0.2591	-0.2 63	-0.4156	0.5454
005	Acenaphthyleno(1,2-b)pentacene	32	39	45.0566	0.3348	0.5196	0.2005	-0.2151	-0.4171	0.4156
006	Acenaphthyleno(1,2-b)hexacene	36	44	50.6691	0.3334	0.4386	0.1505	-0.1668	-0.4176	0.3273
007	Benzo(j)fluoranthene	20	24	28.2118	0.3422	0.6733	0.5253	-0.3116	-0.7689	0.8369
008	Naphtho(2,3-j)fluoranthene	24	29	33.8503	0.3397	0.6795	0.3816	-0.2666	-0.5784	0.6482
009	Acenaphthyleno(1,2-a)naphthacene	28	34	39.4701	0.3374	0.6579	0.2814	-0.2246	-0.4590	0.5060
010	Acenaphthyleno(1,2-a)pentacene	32	39	45.0842	0.3355	0.5629	0.2133	-0.1855	-0.3880	0.3988
011	Acenaphthyleno(1,2-a)hexacene	36	44	50.6965	0.3340	0.4713	0.1661	-0.1518	-0.3414	0.3179
012	Benzo(a)fluoranthene	20	24	28.1636	0.3401	0.7181	0.4514	-0.2521	-0.8444	0.7035
013	Naphtho(2,3-a)fluoranthene	24	29	33.7896	0.3376	0.6986	0.3338	-0.1886	-0.6564	0.5224
014	Indeno(1,2,3-de)pentacene	28	34	39.4059	0.3355	0.6180	0.2523	-0.1481	-0.5031	0.4004
015	Indeno(1,2,3-de)hexacene	32	39	45.0192	0.3338	0.5165	0.1952	-0.1197	-0.4003	0.3149
016	Indeno(1,2,3-de)heptacene	36	44	50.6313	0.3325	0.4341	0.1544	-0.0988	-0.3284	0.2531
017	Benzo(b)fluoranthene	20	24	28.2635	0.3443	0.6874	0.6021	-0.3770	-0.7385	0.9791
018	Naphtho(2,3-b)fluoranthene	24	29	33.9190	0.3420	0.6707	0.4552	-0.3359	-0.5742	0.7910
019	Fluoreno(1,9-ab)naphthacene	28	34	39.5451	0.3396	0.6631	0.3332	-0.2736	-0.4904	0.6067
020	Fluoreno(1,9-ab)pentacene	32	39	45.1616	0.3375	0.6131	0.2499	-0.2154	-0.4424	0.4653
021	Fluoreno(1,9-ab)hexacene	36	44	50.7749	0.3358	0.5191	0.1926	-0.1702	-0.4035	0.3628

81. Cata-condensed Tetracyclic Hydrocarbons

No.	Compound		n	m	W	DE$_{sp}$	k$_2$	k$_1$	k$_{-1}$	k$_{-2}$	E(N→V$_1$)
001	Cyclopenta(e)cyclohept(a)azulene	–	18	21	24.5950	0.3140	0.4745	0.4293	-0.3182	-0.4348	0.7475
002	Cyclopenta(f)cyclohept(a)azulene	–	18	21	24.5219	0.3106	0.6727	0.2607	-0.2451	-0.5849	0.5057
003	Cyclopenta(g)cyclohept(a)azulene	–	18	21	24.5832	0.3135	0.5434	0.3578	-0.3577	-0.4286	0.7155
004	Cyclopenta(e)cyclohept(b)azulene	–	18	21	24.5291	0.3109	0.5946	0.2949	-0.2288	-0.5178	0.5237
005	Dicyclohepta(a,e)pentalene	–	18	21	24.4553	0.3074	0.7621	0.1130	-0.3111	-0.5667	0.4241
006	Dicyclohepta(a,f)pentalene	–	18	21	24.5106	0.3100	0.5354	0.2163	-0.4075	-0.4450	0.6238
007	Dicyclopenta(b,i)heptalene	–	18	21	24.5027	0.3097	0.5210	0.4410	-0.2266	-0.3724	0.6677
008	Dicyclopenta(b,h)heptalene	–	18	21	24.4452	0.3069	0.6180	0.3460	-0.1087	-0.5364	0.4547
009	Dicyclopenta(a,h)heptalene	–	18	21	24.5281	0.3109	0.5377	0.4647	-0.1886	-0.4345	0.6533
010	Dicyclopenta(a,i)heptalene	–	18	21	24.4735	0.3083	0.6180	0.3763	-0.0975	-0.5466	0.4738
011	Dicyclopenta(a,j)heptalene	–	18	21	24.5535	0.3121	0.6180	0.4558	-0.1627	-0.5195	0.6185
012	Dicyclopenta(a,g)heptalene	–	18	21	24.5022	0.3096	0.6856	0.3954	-0.0877	-0.6480	0.4831
013	Dicyclopenta(e,g)azulene	–	16	19	21.4773	0.2883	0.6180	0.3565	0.2466	-0.4731	0.1099
		diA	18	21	21.9704	0.4195	0.3565	0.2466	-0.4731	-1.1899	0.7197
014	Dicyclopenta(a,c)heptalene	–	18	21	24.4348	0.3064	0.5811	0.3326	-0.0926	-0.4782	0.4252
015	Dicyclohept(e,h)azulene	–	20	23	27.0492	0.3065	0.5828	0.1494	-0.3620	-0.4450	0.5113
016	Dicyclohept(e,g)azulene	–	20	23	26.8609	0.2983	0.5733	0.0000	-0.2058	-0.4678	0.2058
		diC	18	23	26.8609	0.3853	0.8853	0.5733	0.0000	-0.2058	0.5733
017	Dicyclohepta(a,c)heptalene	–	22	25	29.2273	0.2891	0.2461	-0.1172	-0.2649	-0.4450	0.1477
		diC	20	25	29.4616	0.3785	1.0287	0.2461	-0.1172	-0.2649	0.3633

82. Peri-condensed Tetracyclic Hydrocarbons

No.	Compound		n	m	W	DE$_{sp}$	k$_2$	k$_1$	k$_{-1}$	k$_{-2}$	E(N→V$_1$)
001	Pyracene	diA	14	15	16.7020	0.4468	0.3820	0.2541	-0.6180	-1.6180	0.8721
		-	12		16.1937	0.2796	0.6180	0.3820	0.2541	-0.6180	0.1279
002	Benz(bc)acepentylenyl	A	14	16	18.1211	0.3826	0.5192	0.3444	-0.2846	-1.2242	0.6290
		C	12		17.4323	0.3395	0.8308	0.5192	0.3444	-0.2846	0.1748
003	Dicyclopenta(cd,hi)indenyl	A	14	16	18.2017	0.3876	0.7485	0.3046	-0.3688	-1.0000	0.6734
		C	12		17.5925	0.3495	0.8323	0.7485	0.3046	-0.3688	0.4440
004	Cyclopenta(cd)cyclohepta-(gh)pentalene	diA	16	17	19.1279	0.4193	0.3521	0.0000	-0.6091	-1.0000	0.6091
		-	14		19.1279	0.3016	0.6227	0.3521	0.0000	-0.6091	0.3521
005	Dicyclopenta(cd,ij)azulene	-	14	17	19.2798	0.3106	1.0000	0.3320	-0.1971	-0.4041	0.5292
006	Pyracyclene	diA	16	17	19.4156	0.4362	0.4142	0.0000	-1.0000	-1.0000	1.0000
		-	14		19.4156	0.3186	1.0000	0.4142	0.0000	-1.0000	0.4142
007	Dibenzo(cd,gh)pentalene	diA	16	17	19.4098	0.4359	0.4559	0.0000	-1.0000	-1.1317	1.0000
		-	14		19.4098	0.3182	0.7046	0.4559	0.0000	-1.0000	0.4559
008	Cyclopent(bc)acenaphthylene	-	14	17	19.4940	0.3232	0.7873	0.4840	-0.0805	-0.8407	0.5645
009	Indeno(5,4,3-cde)azulenyl	A	16	18	20.7608	0.3756	0.4819	0.2105	-0.4010	-1.0000	0.6115
		C	14		20.3398	0.3522	1.0000	0.4819	0.2105	-0.4010	0.2714
010	Benzo(cd)cyclohepta(gh)-pentalenyl	A	16	18	20.7951	0.3775	0.4683	0.2411	-0.5850	-0.7092	0.8261
		C	14		20.3129	0.3507	0.8424	0.4683	0.2411	-0.5850	0.2272
011	Cyclopenta(cd)benz(ij)-azulenyl	A	16	18	20.6043	0.3669	0.5262	0.0572	-0.3373	-0.8555	0.3945
		C	14		20.4899	0.3605	1.0000	0.5262	0.0572	-0.3373	0.4690

82. Peri-condensed Tetracyclic Hydrocarbons (cont. 1)

No.	Compound		n	m	W	DE_{sp}	k_2	k_1	k_{-1}	k_{-2}	$E(N \rightarrow V_1)$
012	Cyclopenta(cd)cyclohept-(hi)indenyl	A	16		20.7611	0.3756	0.5624	0.1799	-0.5413	-0.7152	0.7212
		C	14	18	20.4013	0.3556	0.8071	0.5624	0.1799	-0.5413	0.3825
013	Dicyclopenta(ef,kl)heptalene	-	16	19	22.0515	0.3185	0.5828	0.4450	-0.3620	-0.4450	0.8070
014	Dicyclohepta(cd,gh)pentalene	-	16	19	22.0297	0.3174	0.4944	0.4450	-0.4450	-0.4785	0.8901
015	Cyclopenta(cd)cyclohept(ij)-azulene	-	16	19	21.9485	0.3131	0.6283	0.3029	-0.3072	-0.4880	0.6101
016	Cyclopenta(cd)phenalenyl	A	16		21.0450	0.3914	0.5928	0.2411	-0.7092	-1.0000	0.9503
		C	14	18	20.5628	0.3646	1.0000	0.5928	0.2411	-0.7092	0.3518
017	Benzo(def)fluorenyl	A	16		20.9623	0.3868	0.7691	0.1291	-0.6052	-0.9635	0.7343
		C	14	18	20.7040	0.3724	0.8185	0.7691	0.1291	-0.6052	0.6399
018	Acepleiadylene	-	16	19	22.2517	0.3290	0.6686	0.4450	-0.4450	-0.5030	0.8901
019	Cyclohepta(def)fluorene	-	16	19	21.9929	0.3154	0.7962	0.1761	-0.1323	-0.6672	0.3084
020	Cyclohept(bc)acenaphthylene	-	16	19	22.2063	0.3266	0.8268	0.3291	-0.4048	-0.5041	0.7339
021	Cyclohepta(gh)pleiadiene	-	16	19	22.2197	0.3274	0.6514	0.4109	-0.2815	-0.7287	0.6924
022	Cyclopenta(ef)benzo(kl)-heptalenyl	A	18		22.9284	0.3464	0.6180	-0.2091	-0.3197	-0.7331	0.1107
		C	16	20	23.3465	0.3673	0.7233	0.6180	-0.2091	-0.3197	0.8271
023	Dicyclohept(cd,hi)indenyl	A	18		22.9280	0.3464	0.4631	-0.1812	-0.3490	-0.7101	0.1678
		C	16	20	23.2903	0.3645	0.8877	0.4631	-0.1812	-0.3490	0.6443
024	Indeno(3,4,5-def)heptalenyl	A	18		23.0019	0.3501	0.5847	-0.1531	-0.3931	-0.6515	0.2399
		C	16	20	23.3082	0.3654	0.7239	0.5847	-0.1531	-0.3931	0.7379

82. Peri-condensed Tetracyclic Hydrocarbons (cont. 2)

No.	Compound		n	m	W	DE$_{sp}$	k$_2$	k$_1$	k$_{-1}$	k$_{-2}$	E(N$\rightarrow$V$_1$)
025	Benzo(cd)cyclohept(ij)azulenyl	A	18		23.0775	0.3539	0.3659	-0.0468	-0.3726	-0.7848	0.3258
		C	16	20	23.1712	0.3586	0.9283	0.3659	-0.0468	-0.3726	0.4127
026	Cyclopenta(ef)cyclohept(kl)heptalene	-	18		24.2983	0.2999	0.7654	0.0000	-0.2105	-0.5026	0.2105
		diC	16	21	24.2983	0.3952	0.7708	0.7654	0.0000	-0.2105	0.7654
027	Dicyclohept(cd,ij)azulene	-	18	21	24.4011	0.3048	0.4639	0.1331	-0.2149	-0.7024	0.3481
028	Pyrene	-	16	19	22.5055	0.3424	0.8794	0.4450	-0.4450	-0.8794	0.8901
029	Cyclohepta(cd)phenalenyl	C	16		23.5727	0.3786	1.0000	0.6180	-0.2091	-0.4388	0.8271
		A	18	20	23.1546	0.3577	0.6180	-0.2091	-0.4388	-1.0000	0.2297
030	Cyclohepta(def)phenanthryl	C	16		23.5257	0.3763	0.9303	0.5692	-0.0980	-0.6467	0.6672
		A	18	20	23.3297	0.3665	0.5692	-0.0980	-0.6467	-0.7426	0.5487
031	Cyclohepta(cd)pleiadiene	-	18		24.5328	0.3111	0.7654	0.0000	-0.2873	-0.6889	0.2873
		diC	16	21	24.5328	0.4063	1.0000	0.7654	0.0000	-0.2873	0.7654
032	Dibenzo(ef,kl)heptalene	-	18		24.5368	0.3113	0.7654	0.0000	-0.3011	-0.5767	0.3011
		diC	16	21	24.5368	0.4065	1.0000	0.7654	0.0000	-0.3011	0.7654
033	Cyclohepta(gh)pleiadiene	-	18	21	24.6084	0.3147	0.7133	0.0571	-0.3331	-0.6242	0.3901
034	Benzo(ef)cyclohepta(kl)heptalenyl	C	18		25.7697	0.3532	0.8915	0.1845	-0.2087	-0.3836	0.3933
		A	20	22	25.3523	0.3342	0.1845	-0.2087	-0.3836	-0.5473	0.1749
035	3,4-Cycloheptabenzo(ef)heptalenyl	C	18		25.8185	0.3554	0.8262	0.2263	-0.1997	-0.5343	0.4260
		A	20	22	25.4191	0.3372	0.2263	-0.1997	-0.5343	-0.5485	0.3346
036	Dicyclohepta(ef,kl)heptalene	-	20		26.5523	0.2849	0.3473	-0.1689	-0.2232	-0.3473	0.0543
		diC	18	23	26.8901	0.3865	1.0000	0.3473	-0.1689	-0.2232	0.5162

T